1948 AND AFTER

1948 AND AFTER

ISRAEL AND THE PALESTINIANS

BENNY MORRIS

CLARENDON PRESS · OXFORD

1990

Oxford University Press, Walton Street, Oxford OX2 6DP

Oxford New York Toronto
Delhi Bombay Calcutta Madras Karachi
Petaling Jaya Singapore Hong Kong Tokyo
Nairobi Dar es Salaam Cape Town
Melbourne Auckland

and associated companies in
Berlin Ibadan

Oxford is a trade mark of Oxford University Press

Published in the United States
by Oxford University Press, New York

British Library Cataloguing in Publication Data
Morris, Benny 1948–
1948 and after: Israel and the Palestinians.
1. Israel. Settlement by Palestinian Arabs
I. Title
956.940049274
ISBN 0–19–828784–4

Library of Congress Cataloging in Publication Data
Morris, Benny, 1948–
1948 and after: Israel and the Palestinians / Benny Morris.
Includes index.
1. Israel–Arab War, 1948–1949—Refugees. 2. Refugees, Arab.
3. Israel—Historiography. 4. Palestinian Arabs—Israel.
I. Title.
DS126.954.M67 1990 956.04'2—dc20 90–7469
ISBN 0–19–828784–4

Typeset by Hope Services (Abingdon) Ltd
Printed and bound in Great Britain by
Biddles Ltd, Guildford and King's Lynn

For Leah

Preface

This collection of essays serves as a complement to my study *The Birth of the Palestinian Refugee Problem, 1947–1949*, which was published in 1988. In part, these essays are based on material that I discovered or became available only after completing *The Birth*. They elucidate aspects of the subject not treated or only fleetingly referred to in the original work. They will, I feel, provide the reader with additional insights into, as well as facts about, the Arab exodus from Palestine. The perception of the variety and tragic richness of that revolutionary event is thus enhanced.

Inevitably, since both volumes deal with the exodus, there is a degree of overlap. But I have striven to keep that overlap to a minimum and I feel confident that the readers will not feel it, except perhaps in the essay 'Yosef Weitz and the Transfer Committees, 1948–1949'.

The initial essay, 'The New Historiography: Israel and its Past', is in great measure introductory. It places the collection (as well as *The Birth*) in its historiographic context and summarizes the issues and the arguments that have characterized the controversy sparked in Israel by the publication of *The Birth* and other 'new' histories, most notably Avi Shlaim's *Collusion Across the Jordan* (Oxford, 1988). The reader perhaps will find in some of the essays, particularly in 'Yosef Weitz and the Transfer Committees, 1948–1949' and 'The Causes and Character of the Arab Exodus from Palestine: the Israel Defence Forces Intelligence Service Analysis of June 1948', answers to some of the contentions voiced by the critics of the New Historiography. He will also find, as in 'The Initial Absorption of the Palestinian Refugees in the Arab Host Countries, 1948–1949', explorations and passages that do not conform to the image of the New Historiography that these critics have tried to project.

The essays 'Mapai, Mapam, and the Arab Problem in 1948', 'The Causes and Character of the Arab Exodus from Palestine: the Israel Defence Forces Intelligence Service Analysis of June 1948', and 'Yosef Weitz and the Transfer Committees, 1948–1949' all deal with the core questions surrounding the exodus. 'Mapai, Mapam . . .' traces what was certainly the only major ideological-ethical debate that took place in and near the corridors of Israeli power in 1948 on the question of policy towards the Arabs during the war—and an interesting debate it was. 'The Causes . . .' analyses a unique (and, in my eyes, a historiographically decisive) intelligence document— perhaps one of the few occasions in historiography in which an intelligence document actually clarifies a major historical happening. 'Yosef Weitz . . .' traces in minute detail, almost day by day, the activities of an outstanding Zionist official during the agrarian-demographic revolution of 1948, highlighting how he affected it.

The essays 'Haifa's Arabs: Displacement and Concentration, July 1948', 'The Case of Abu Ghosh and Beit Naqquba, Al Fureidis and Khirbet Jisr az Zarka in 1948—or Why Four Villages Remained', and 'The Transfer of Al Majdal's Remaining Arabs to Gaza, 1950' describe the varying fates of different Arab communities during the 1948 war and in the immediate post-war period. The three 'tales' underline the variety and, often, the highly local character of what happened—despite the shared common denominators and patterns of the exodus in most places at most times. They also shed a good deal of light on the nature of Israeli decision-making. The first two highlight the interplay (and divergence) between Israel's new civilian and military authorities and the increasing restraints imposed on the IDF's power *vis-à-vis* Israel's Arab minority as the hostilities in each sector drew to a close. The third essay, on Majdal (Ashkelon), casts some light on the thinking and manner of operation of one of Israel's most interesting politician-generals, Moshe Dayan, while elucidating the tactics of transfer once the revolutionary stage of Israeli nation-building had ceased.

Lastly, 'The Initial Absorption . . .' sheds light on what happened to the exiles once they moved to the areas under Arab rule and why they became permanent refugees.

Taken together, these essays hope to contribute to an under-standing of the origins of the Palestinian refugee problem and of the early history of Israel's Arab minority.

ACKNOWLEDGEMENTS

I would like to thank the publishers and editors of the following journals for their permission to republish the named essays:

Tikkun, where an early version of 'The New Historiography: Israel and its Past' first appeared in 1988; and the directors of Georgetown University's Center for Contemporary Arab Studies, where I delivered an earlier version of the essay as a paper at a conference in 1989.

Middle Eastern Studies, which in 1986 published 'The Causes and Character of the Arab Exodus from Palestine: The IDF Intelligence Service Analysis of June 1948' and 'Yosef Weitz and the Transfer Committees, 1948–1949'.

The Middle East Journal, which in 1986 published 'The Harvest of 1948 and the Creation of the Palestinian Refugee Problem' and, in 1988, 'Haifa's Arabs: Displacement and Concentration, July 1948'.

And the publishers Unwin Hyman, who first published 'The Initial Absorption of the Palestinian Refugees in the Arab Host Countries 1948–1949' in their volume *Refugees in the Age of Total War* (1988).

I would also like to thank Henry Hardy and the editors of OUP for seeing this collection over the hurdles of publication.

CONTENTS

LIST OF MAPS

ABBREVIATIONS

AFSC	American Friends Service Committee Archive, Philadelphia
CP	Sir Alan Cunningham Papers, Middle East Centre Archives, St Antony's College, Oxford
CZA	Central Zionist Archives, Jerusalem
DBG	David Ben-Gurion's *Yoman Hamilhama-Tashah* (the war diary 1948–9)
HHA	Hashomer Hatza'ir Archives (Mapam, Kibbutz Artzi Papers), Givat Haviva, Israel
HHA-ACP	Hashomer Hatza'ir Archives, Aharon Cohen Papers, Givat Haviva, Israel
HMA	Haifa Municipal Archives, Haifa Town Hall
IDF	Israel Defence Forces
ISA	Israel State Archives, Jerusalem
AM	Agriculture Ministry Papers
FM	Foreign Ministry Papers
JM	Justice Ministry Papers
MAM	Minority Affairs Ministry Papers
PMO	Prime Minister's Office Papers
IZL	Irgun Zva'i Leumi (National Military Organization) or the Irgun
JI	Jabotinsky Institute (IZL and LHI Papers), Tel Aviv
JEM	Jerusalem and East Mission Papers, Middle East Centre Archives, St Antony's College, Oxford
JNF	Jewish National Fund
KMA	Kibbutz Meuhad Archives ⎫
KMA-AZP	Kibbutz Meuhad Archives, Aharon Zisling Papers ⎬ Ef'al, Israel

KMA-PA	Kibbutz Meuhad Archives, Palmah Papers } Ef'al, Israel
LA	Labour Archives (Histadrut), Lavon Institute, Tel Aviv
LHI	Lohamei Herut Yisrael (Freedom Fighters of Israel) or Stern Gang
LPA	Labour Party Archives (Mapai), Beit Berl, Tzofit, Israel
MAC	Mixed Armistice Commission
NA	National Archives, Washington, DC
PRO	Public Record Office, London
CAB	Cabinet Office Papers
CO	Colonial Office Papers
FO	Foreign Office Papers
WO	War Office Papers
UN	United Nations Archives, New York City

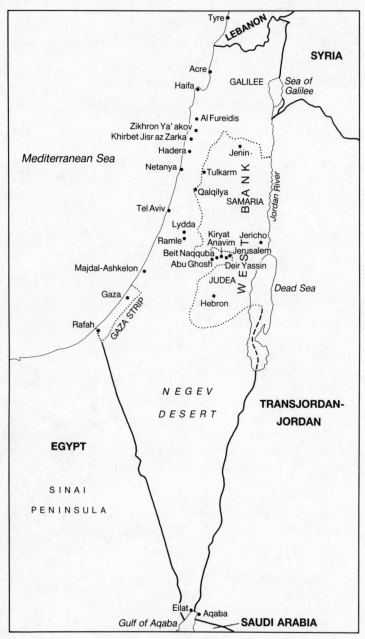

FIG. 1. Israel/Palestine, showing 1949 Armistice frontiers

1

The New Historiography: Israel and its Past

On 11 July 1948, the Yiftah Brigade's Third Battalion, as part of Operation Dani, occupied the Arab town of Lydda. The battalion's 400–500 troops, supported by a company from the First Battalion, took up positions around the centre of the town; the remaining population, some 30,000 Arabs, shut themselves up in their homes. There was no formal surrender, but the night passed quietly. Just before noon the following day, two or three gun-mounted armoured cars belonging to the Arab Legion, the British-led Transjordanian army, drove into town. They were either on a reconnaissance mission or were unaware that the town had already fallen into Israeli hands. A firefight ensued, and the scout cars withdrew. But a number of armed townspeople, perhaps believing that the shooting heralded a major Arab counter-attack, began sniping from windows and rooftops at their Israeli occupiers. The Third Battalion—several hundred nervous soldiers in the middle of an Arab town of tens of thousands—fiercely put down what various chroniclers, Jewish and Arab, subsequently called a 'rebellion', firing in the streets, into houses, and at the concentrations of POWs in the mosque courtyards. Israeli military records refer to 'more than 250' Arabs killed in the town that afternoon. By contrast, Israeli casualties in both the firefight with the armoured cars and in the suppression of the sniping were between two and four dead (the reports vary) and twelve or so wounded.

Israeli historians called the affair a 'rebellion' in order to justify the slaughter and the subsequent expulsion; Arab chroniclers, such as Aref al Aref, did likewise in order to emphasize Palestinian resolve and sacrifice in the battle against the Zionist invaders.

Operation Dani took place roughly midway through the first Israeli–Arab war, Israel's War of Independence. The Arab states' invasion of the fledgling Jewish state on 15 May had been halted weeks before; the newly organized and freshly equipped Israel Defence Forces (IDF) were on the offensive on all fronts, which was to be the case for the remainder of the war.

At 13.30 hours on 12 July, before the shooting in Lydda had completely died down, Lieutenant-Colonel Yitzhak Rabin, Operation Dani head of operations, issued the following order: '1. The inhabitants of Lydda must be expelled quickly without attention to age. They should be directed towards Beit Nabala. Yiftah [Brigade HQ] must determine the method and inform [Operation] Dani HQ and 8th Brigade HQ. 2. Implement immediately.' A similar order was issued at the same time to the Kiryati Brigade concerning the inhabitants of the neighbouring Arab town of Ramle, occupied by Kiryati troops that morning. There had been no shooting in Ramle that day. Ramle's notables had signed a formal surrender instrument a few hours before; no such surrender had taken place in Lydda.

On 12 and 13 July, the Yiftah and Kiryati brigades carried out their orders, expelling the 50–60,000 remaining inhabitants of and refugees camped in and around the two towns, which lie some 10 miles south-east of Tel Aviv. Throughout the war, the two towns, which sat astride the main Tel Aviv–Jerusalem road, had interdicted Jewish traffic. Moreover, the leaders of the Yishuv, Palestine's Jewish community, regarded Lydda and Ramle, at first held by Arab irregulars and, from May, by the Arab Legion, as a perpetual threat to Tel Aviv itself, a springboard from which the Arabs could attack the Jewish capital.

About noon on 13 July, Operation Dani HQ informed IDF General Staff / Operations: 'Lydda police fort has been captured. [The troops] are busy expelling the inhabitants [*oskim begeirush hatoshavim*].' Lydda's inhabitants were forced to walk eastwards to the Arab Legion lines; many of Ramle's inhabitants were ferried in trucks or buses. Clogging the roads (and the Legion's possible routes of advance westward), the tens of thousands of refugees marched, gradually shedding their worldly goods along the way. It was a hot summer day. Arab chroniclers, such as Sheikh Muhammad Nimr al Khatib, claimed that hundreds of children died in the march, from dehydration and disease. One Israeli witness described the spoor: the refugee column 'to begin with [jettisoned] utensils and furniture and, in the end, bodies of men, women and children . . .'. Many of the refugees came to rest near the town of Ramallah and set up tent encampments, which became some of today's mortar and concrete-block West Bank refugee camps.[1]

[1] A full description of the Lydda–Ramle episode is to be found in Benny Morris,

Israeli historians during the 1950s, 1960s, and 1970s were less than candid in their treatment of the Lydda–Ramle episode.

The IDF's official history of the war, *Toldot Milhemet Hako-memiyut* (History of the War of Independence), produced by the General Staff / History Branch and published in 1959, stated: 'The Arabs [of Lydda], who had violated the terms of the surrender and feared [Israeli] retribution, were happy at the possibility given them of evacuating the town and proceeding eastwards, to Legion territory; Lydda emptied of its Arab inhabitants.'[2] The official history makes no reference to the exodus of the inhabitants of Ramle or to its reasons.

A few years later, in 1961, the former head of the IDF General Staff / History Branch, Lieutenant-Colonel (res.) Natanel Lorch, published his history of the war, *The Edge of the Sword*, a slightly expanded version of *Toldot Milhemet Hakomemiyut*. In it he explained the exodus from Lydda thus: 'The residents, who had violated surrender terms and feared retribution, declared they would leave and asked for safe conduct to Arab Legion lines, which was granted.' The same sentence appears in the revised edition of *The Edge of the Sword*, published by Lorch in 1968.[3]

A somewhat less deceitful, but still misleading, description of the events in Lydda and Ramle is provided by Lieutenant-Colonel (res.) Elhannan Orren, another former officer of the IDF History Branch, in his *Baderekh el Ha'ir* (On the Road to the City), a highly detailed history of Operation Dani published by the IDF press, Ma'arachot, in 1976. Like his predecessors, Orren failed to state anywhere that what had occurred was an expulsion, and one explicitly ordered from on high. He writes:

When the police fort fell [on 13 July]—the morale of the inhabitants fell.... The [Lydda] notables asked that they be allowed to take with them the incarcerated menfolk, and then they would leave without delay. Their request was granted . . . and a mass evacuation of Lydda began. . . . The announcement about the exodus from Lydda was transmitted to Kiryati in

'Operation Dani and the Palestinian Exodus from Lydda and Ramle in 1948', *The Middle East Journal*, 40, 1, (Winter, 1986), 82–109.

[2] IDF General Staff / History Branch, *Toldot Milhemet Hakomemiyut* (Tel Aviv, 1959), 259.

[3] N. Lorch, *The Edge of the Sword* (Tel Aviv, 1961), 286, and revised edn. 1968, 340.

Ramle. Here, transport was provided, and the troops made [the evacuation] easier for the inhabitants. The stream of departure, which began already before the conquest, grew on the 12 of July.[4]

(In fact, the expulsion order had originated with Ben-Gurion, according to the Prime Minister's first major biographer, Michael Bar-Zohar, and most recent Israeli histories, such as Arieh Itzchaki's *Latrun*).[5]

Ironically, a franker description of the episode was provided by Yitzhak Rabin, who stated in his autobiography, *Pinkas Sherut* (in English, *The Rabin Memoirs*), that what had occurred in Lydda and Ramle had been an expulsion. But the relevant passage was excised by Israeli government censors. (Subsequently, to the retired Prime Minister's embarrassment, the offending passage was published in the *New York Times* by Rabin's translator into English, Peretz Kidron.)[6]

The treatment of the Lydda–Ramle affair by past Israeli historians is illustrative of what can be called, for want of a better term, the 'old' or 'official' history. That history has shaped the way Israelis and Diaspora Jews—or, at least, Diaspora Zionists—have seen and, in large measure, still see Israel's past; and it has also held sway over the way gentile Europeans and Americans (and their governments) see that past. This understanding of the past has significantly influenced the attitudes of Diaspora Jews, as well as the attitudes of European and American non-Jews, toward Israel over the decades, affecting government policies concerning the Israeli–Arab conflict.

The essence of the old history is that Zionism's birth was an inevitable result of Gentile pressures and persecution, and that it offered at least a partial solution to the 'Jewish Problem' in Europe; that the Zionists intended no ill to the Arabs of Palestine, and that Zionist settlement alongside the Arabs did not, from the Jews' point of view, necessitate a clash or displacement, but that Israel was born

---

[4] E. Orren, *Baderekh el Ha'ir*, [on the road to the city] (Tel Aviv, 1976), 124.

[5] Michael Bar-Zohar, *Ben-Gurion* (Tel Aviv, 1977), ii. 775; Arieh Itzchaki, *Latrun*, ii, 394.

[6] *The New York Times*, 23 Oct. 1979. The first Israeli historian to my knowledge to treat the Lydda–Ramle episode with relative candour, if rather briefly, was Arieh Itzchaki, in his 2-volume *Latrun, the Battle for the Road to Jerusalem* (Tel Aviv, 1982), 392–4. But he too speaks of 'a general rebellion' of the Lydda townspeople; of the townspeople's 'treachery', as they had 'signed a surrender' instrument before this; and of the local dignitaries proposing the evacuation of the town.

into an uncharitable, predatory environment; that Zionist efforts at compromise and conciliation were rejected by the Arabs; and that the Palestinians and the neighbouring Arab states, selfish and ignoble, refused to accede to the burgeoning Zionist presence and in 1947–8 attacked the Yishuv with the aim of nipping the Jewish state in the bud. The Arabs, so goes the old history, were far stronger politically and militarily than the Yishuv and were assisted in their efforts by the British, but none the less lost the war. The Haganah at first defeated the *knufiyot* (or 'gangs', the Yishuv's name for the Arab irregulars and militiamen), who were reinforced by contingents of foreign Arab volunteers (the Arab Liberation Army), and then the five (Egypt, Syria, Iraq, Transjordan, and Lebanon) invading Arab armies. In the course of that war, says the old history, in order to facilitate the invasion of the Arab armies, the Arab leaders called upon / ordered Palestine's Arabs to quit their homes: this would lay the Jewish state open to charges of expulsion and physically clear the path for the Arab armies. Thus was born the Palestinian refugee problem.

The old history—whose purest expression, incidentally, is to be found in Israeli elementary- and high-school history textbooks and in David Ben-Gurion's own 'histories' of the Yishuv and Israel's first years (*Medinat Yisrael Hamehudeshet, Behilahem Yisrael*, and so on)—generally makes the claim that in the immediate wake of the 1948 war, Israel's leaders desperately sought to make peace with one or all of the Arab states but the Arabs, hell-bent on the destruction of Israel, turned down these overtures and repaid this goodwill with continued belligerence.

The old historians, who perhaps should more accurately be called chroniclers, offered a simplistic and consciously pro-Israeli interpretation of the past, and generally avoided mention of anything that reflected poorly upon Israel. It was argued, with a measure of justice, that since the conflict with the Arabs was still raging, and since it was a political as well as military struggle, it necessarily involved propaganda, the good (or ill) will of governments in the West, and the hearts and minds of Christians and Diaspora Jews. Blackening Israel's name, it was argued, would ultimately weaken Israel in its ongoing struggle for survival. *Raisons d'état* often took precedence over telling the truth.

The past few years have witnessed the emergence of a new

generation of Israeli scholars and a 'new' history. These historians, some of them living or working abroad, have looked and are looking afresh at the Israeli historical experience, and their conclusions are often at odds with those of the old history.

Some critics have recently called the 'new' history 'revisionist'. I don't like the name for three reasons. First, it conjures up the faces of Ze'ev Jabotinsky and Menachem Begin, respectively the founder / prophet and latter-day leader of the Revisionist Movement in Zionism. The Revisionists, unpragmatic, right-wing deviants from the mainstream of the Zionist experience, claimed all of Palestine and the East Bank of the Jordan River for the Jews. Until 1977, their place was on the fringe of Zionist and Israeli history, though since then their vision, albeit in diluted form, has dominated the political arena in Jerusalem. To call the new history 'revisionism' is to cause unnecessary confusion and, to some, anguish.

Secondly, in western historiography, revisionism recently has denoted those, like A. J. P. Taylor, who have looked anew at the origins of either of the two world wars or both and have concluded that the blame for their outbreak must be distributed more equitably as between the Germans and the Allied Powers. And, in a minor branch of this genus, there are some who, to dilute German guilt, have even denied any Nazi wrongdoing or the Holocaust itself. A more recent wave of revisionists have sought to lessen Soviet blame for the Cold War. On this count, too, the term 'revisionism' in connection with Israeli historiography can only mislead and confuse.

Lastly, the term 'revisionism' would seem to imply that there already existed a solid, credible—if wrong-headed—body of historiography which latest fashion is bent on overthrowing, perhaps for ulterior reasons. But this is not really the case. Israel's old historians, by and large, were not really historians and did not produce real history. In reality, they were chroniclers, and often apologetic, interested chroniclers at that. They did not work from and upon a solid body of contemporary documentation and did not normally try to paint a picture that offered the variety of sides of a given historical experience. They worked from interviews and memoirs, and often from memories. They had neither the mind-set nor the materials to write real history. To call the new history 'revisionist' would, by implication, ascribe too much merit to the old history.

Two factors apply in the emergence of the new history, one relating to materials and the other to personae.

Thanks to Israel's Archives Law (passed in 1955 and amended in 1964 and again in 1981), and to its relatively liberal implementation, from the early 1980s onwards hundreds of thousands, perhaps millions, of state documents were opened to researchers. Almost all the Foreign Ministry's papers from 1948 to 1957, as well as great masses of documents—memoranda, minutes, correspondence—from other ministries, including the Prime Minister's Office (though largely excluding the papers of the Defence Ministry and the IDF), have been declassified. Similarly, the past few years have seen the organization and release of large collections of private papers and political party papers.

For the first time, therefore, historians—Israelis and non-Israelis—have been able to write studies about the first years of Israeli statehood and the first years of the Israeli–Arab conflict on the basis of a large collection of contemporary source-material.

The second factor is the nature of the new historians. Most of them, born around 1948, have matured in a more open, doubting, and self-critical Israel than the pre-1967, pre-1973, and pre-Lebanon-War Israel of the old historians. Most of the old historians, indeed, had lived through 1948 as highly committed adult participants in the epic, glorious rebirth of the Jewish commonwealth. They were unable to separate their lives from the events they later recounted, unable to distance themselves from and regard impartially the facts and processes through which they had lived. Most, if asked, will admit as much—and in this, incidentally, they share a common experience with most chronicler-participants in national liberation movements and revolutions in the modern age. The generation of nation-builders seldom casts doubt or looks frowningly back upon its handiwork.

By contrast, the new generation of Israeli historians have been capable of at least a measure of impartiality, and of sometimes more than that. Inevitably, the new historians have focused their attention on the 1947–51 period, both because the documents were available and because that first Israeli–Arab war and its aftermath were the central, natal, revolutionary event in Israeli and, perhaps, modern Jewish history. How one perceives 1948 bears heavily upon how one perceives, interprets, and understands the whole of the

Zionist/Israeli experience. For some, 1948 carries the brand of
original sin; for others, it was a period marked by ineluctable
savagery, but also by innocence, courage, and glory.

The past few years have seen the publication in the West and Israel
of a handful of 'new' histories, including Avi Shlaim's *Collusion
Across the Jordan* (Oxford, 1988); Uri Bar-Joseph's *The Best of
Enemies, Israel and Transjordan in the War of 1948* (London, 1987);
Ilan Pappe's *Britain and the Arab–Israeli Conflict, 1948–1951*
(London, 1988); Simha Flapan's *The Birth of Israel* (New York,
1987)—not strictly a work of history, which I shall deal with
presently; Yitzhak Levy's *Tish'ah Kabin* [Jerusalem in the War
of Independence] (IDF Press, 1986); Yossi Amitai's *Ahvat Amim
Bemivhan* [Testing Brotherhood] (Tel Aviv, 1988); Michael Cohen's
Palestine and the Great Powers, 1945–1948 (Princeton University
Press, 1982); Anita Shapira's *Me'Piturei Haram'a ad Peiruk
Ha'Palmah* [The Army Controversy, 1948] (Tel Aviv, 1985) and
Ha'Halicha al Kav Ha'Ofek [Visions in Conflict] (Tel Aviv, 1989);
and my own *The Birth of the Palestinian Refugee Problem, 1947–
1949* (Cambridge, 1988). A monumental, if controversial, work of
new history, Uri Milstein's *History of the War of Independence*, has
begun to appear; so far three of its projected twelve volumes have
been published (as of April 1989).

To these I would add, though slightly anticipating the 1980s,
Yehoshua Porath's excellent studies of the origins of Palestinian
nationalism, *The Emergence of the Palestinian–Arab National
Movement 1918–1929* and *The Palestinian Arab National Movement
1929–1939* (London, 1974 and 1979).

Taken together, these works—along with a large number of
articles that have appeared recently in academic journals such as
Middle Eastern Studies, *Studies in Zionism*, and *The Middle East
Journal*—significantly undermine, if not thoroughly demolish, a
variety of assumptions that helped form the core of the old history.

Flapan's *The Birth of Israel* is the least historical of these books.
Indeed, it is not, strictly speaking, a history at all but rather a polemical
work written from a Marxist perspective. In his introduction,
Flapan—the former head of Mapam's Arab Department who passed
away in 1987—wrote that his purpose was not to produce 'a detailed
historical study interesting only to historians and researchers', but
rather to write 'a book that will undermine the propaganda structures

that have so long obstructed the growth of the peace forces in my country . . .' Politics rather than historiography is the book's manifest objective.

But despite this explicitly polemical purpose, Flapan's book has the virtue of more-or-less accurately formulating some of the central fallacies—which he calls 'myths'—that inform the old history. These were: (1) that the Yishuv in 1947 joyously embraced partition and the truncated Jewish state prescribed by the UN General Assembly, and that the Palestinians and the surrounding Arab states unanimously rejected the partition and attacked the Yishuv with the aim of throwing the Jews into the sea; (2) that the war was waged between a relatively defenceless and weak (Jewish) David and a relatively strong (Arab) Goliath; (3) that the Palestinians fled their homes and villages either 'voluntarily'—meaning for no good reason—or at the behest / orders of the Arab leaders; and (4) that, at the war's end, Israel was interested in making peace, but the recalcitrant Arabs preferred a perpetual war to the finish.

PARTITION AND WAR AIMS

Because of poor research and analysis, Flapan's demolition of these myths is far from convincing. Shlaim, in *Collusion Across the Jordan*, tackles some of the myths more persuasively. The original goal of Zionism was the establishment of a Jewish state in the whole of Palestine. The acceptance of partition, in the mid-1930s as in 1947, was tactical, not a change in the Zionist dream. 'I don't regard a state in part of Palestine as the final aim of Zionism, but as a means towards that aim', Ben-Gurion wrote in 1938.[7] A few months earlier, Ben-Gurion told the Jewish Agency Executive that he supported partition 'on the basis of the assumption that after we constitute a large force following the establishment of the state—we will cancel the partition of the country and we will expand throughout the Land of Israel.'[8] To his wife, Paula, Ben-Gurion wrote: 'Establish a Jewish state at once, even if it is not in the whole land. The rest will come in the course of time. It must come.'[9]

[7] Quoted in Shabtai Teveth, *Ben-Gurion and the Palestine Arabs* (Oxford, 1985), 188.
[8] Quoted in Benny Morris, *The Birth of the Palestinian Refugee Problem, 1947– 1949* (Cambridge, 1988), 24.
[9] Quoted in Avi Shlaim, *Collusion Across the Jordan* (Oxford, 1988), 17.

The Yishuv entered the first stage of the war in November–December 1947 with an understanding with Transjordan's King Abdullah—'a falcon trapped in a canary's cage'—that, come the British evacuation, his army, the Arab Legion, would take over the eastern part of Palestine (now called the West Bank), earmarked by the UN to be the core of the Palestinian Arab state, and that it would leave the Yishuv alone to set up the Jewish state in the other areas of the country. The Yishuv and the Hashemite Kingdom of Transjordan, Shlaim and Bar-Joseph persuasively argue, had conspired from 1946 to early 1948 to nip the impending UN partition resolution in the bud and to thwart the emergence of a Palestinian Arab state. It was to be partition, but between Israel and Transjordan. This 'collusion' and 'unholy alliance'—in Shlaim's loaded phrases —was sealed at the now-famous clandestine meeting between Golda Myerson (Meir) and Abdullah at Naharayim on the Jordan River on 17 November 1947.[10]

Moreover, this Zionist–Hashemite understanding was fully sanctioned by the British Government, Shlaim, Milstein, Bar-Joseph, and Pappe demonstrate. Contrary to the old Zionist historiography, which was based largely on the mistaken feelings of Israel's leaders at the time, Britain's Foreign Secretary, Ernest Bevin, by February 1948 had clearly 'become resigned to the inevitable emergence of a Jewish state', while opposing the emergence of a Palestinian Arab state. And he explicitly warned Transjordan 'to refrain from invading the areas allotted to the Jews'.[11]

Both Shlaim and Flapan make the point that the Palestinian Arabs, though led by Haj Amin al Husayni, the conniving, extremist, former mufti of Jerusalem, were far from unanimous in supporting the Husayni-led crusade against the Jews. Indeed, in the first months of the hostilities, according to the Yishuv's intelligence sources, the bulk of Palestine's Arabs merely wanted peace and quiet, if only out of a healthy respect for the Jews' martial prowess. But gradually, in part because of Haganah over-reactions, the conflict spread, eventually engulfing the two communities throughout the land. In April and May 1948, the Haganah gained the upper

---

[10] Shlaim, *Collusion*, pp. 110–16; and Ilan Pappe, *Britain and the Arab–Israeli Conflict, 1948–1951* (Macmillan, 1988), 10.
[11] Shlaim, *Collusion*, p. 139; and Pappe, *Britain and the Arab–Israeli Conflict*, pp. 16–18.

hand and the Palestinians lost the war, most of them going into exile.

What ensued, once Israel declared its independence on 14 May 1948 and the Arab states invaded on 15 May, was 'a general land grab', with everyone—Israel, Transjordan, Syria, Egypt, and even Lebanon—bent on preventing the birth of a Palestinian Arab state and carving out chunks of Palestine for themselves.

Contrary to the old historiography, Abdullah's invasion of eastern Palestine was clearly designed to conquer territory for his kingdom —at the expense of the Palestinian Arabs—rather than to destroy the Jewish state. Indeed, the Arab Legion stuck meticulously, throughout the war, to its non-aggressive stance *vis-à-vis* the Yishuv and the Jewish state's territory (in part, in line with the November 1947 'understanding' and, in part, out of reluctance to commit the army in a battle in which it would be mauled. Abdullah understood that the Legion was his Kingdom's shield; without it, he would fall.)

There were two exceptions: the Legion's support for the Palestinian irregulars in overrunning the Etzion Bloc of Jewish settlements, south of Bethlehem, and the Legion's capture of the Jewish Quarter of the Old City and the probes towards West Jerusalem, at Notre Dame and the Mandelbaum Gate, in the second half of May. But the Etzion Bloc was an island of Jewish settlement inside the territory earmarked for Arab sovereignty, an island that interdicted Arab traffic along the vital Hebron–Bethlehem–Jerusalem road. And Jerusalem was an area earmarked for international rule rather than Jewish or Arab sovereignty. Jerusalem was never covered by the Yishuv–Hashemite 'understanding' and was, effectively, up for grabs throughout the war. Yet Arab Legion commander Glubb Pasha for days resisted local Palestinian importuning to advance into Jerusalem, for fear of losing his small army in costly and indecisive street battles. He moved in, reluctantly, on 17–18 May, and only after desperate appeals from the local Arab population had forced Abdullah's hand. The Legion's pushes at Notre Dame and the Mandelbaum Gate were apparently designed to relieve pressure, or expected pressure, on the Old City and Arab East Jerusalem in general, and were never pressed with determination. Conquest of Jewish West Jerusalem was never, and was never seen to be, on the cards.

Rather, in the course of May–July 1948, it was the Haganah / IDF
that violated the November 'understanding' and repeatedly assaulted
the Arab Legion in areas earmarked by the partition resolution for
Arab sovereignty—at Latrun, Lydda–Ramle, Nabi Samwil, and so
on. Abdullah and Glubb, from the start, had correctly assessed
Jewish determination and military strength and had realized that
their army was barely capable of seizing and controlling Arab
Eastern Palestine let alone having strength to spare for offensive
operations against the Jews and against Israeli territory. It is not at
all clear that Abdullah and Glubb would have been happy to see
the collapse in May 1948 of the fledgling Jewish republic. Certainly
Abdullah was far more troubled by the prospects of the emergence
of a Palestinian Arab state and of an expanded Syria and an
expanded Egypt on his frontiers than by the emergence of a small
Jewish state. Abdullah and Glubb desisted from an offensive strategy
against the embattled Jewish state during the crucial 15 May–11
June period. Abdullah neither expected nor planned for Israel's
demise; nor were the Arab Legion's operations from May onwards
designed to bring this about.[12]

As to the other Arab states' hopes, objectives, plans, roles, and
operations during the invasion of Palestine / Israel, Shlaim and
Pappe are not completely clear. And, so long as the Arab states'
archives remain closed to researchers, it is doubtful whether a clear,
definitive conclusion on these matters is at all possible. Was the
primary aim of Egypt, Iraq, Syria, and Lebanon in invading to
overrun the Yishuv and destroy the Jewish state or was it merely or
mainly to frustrate or curtail Abdullah's territorial ambitions and to
acquire some territory for themselves? Or did they have both these
aims—destroying Israel and frustrating Abdullah—in mind? Or
were they motivated to invade mainly by a desire to appease
internal pressures or expected internal pressures to come to the aid
of the Palestinian brothers and to fend off threats to their own
regimes stemming from this pro-Palestinian ground-swell?

Flapan argues firmly, but without evidence, that 'the invasion . . .
was not aimed at destroying the Jewish state'. Shlaim and Pappe are
more cautious. Shlaim writes that the Arab armies intended to
bisect the Jewish state and, if possible, 'occupy Haifa and Tel Aviv'

[12] Uri Bar-Joseph, *The Best of Enemies, Israel and Transjordan in the War of 1948*
(London, 1987), 240–5.

or 'crippl[e] the Jewish state'. Here, manifestly, we are talking about destruction. But Shlaim at the same time argues that the Arab states were driven to the invasion more by a desire to stymie Abdullah than by a wish to kill the Jews; and, partly for this reason, they did not properly plan the invasion. None of the Arab states, save Transjordan, committed the full weight of their military power to the enterprise—indicating either inefficiency or, perhaps, a less than wholehearted seriousness about the declared aim of driving the Jews into the sea. In any case, Transjordan frustrated the other Arabs' plans and intentions, and rendered their military preparations and planning ineffective.[13]

The Egyptian government during the months before May 1948 had consistently opposed participation in an invasion of Palestine / Israel. The army commanders maintained that the army was 'ill-equipped and ill-trained for such a formidable operation'. The public apparently felt otherwise, expecting the pan-Arab invasion of the Zionist entity to be a walkover. King Farouq eventually prevailed over the generals and his Prime Minister, Muhammad Fahmi an Nuqrashi Pasha, and ordered the army to march. Farouq 'probably hoped to be able to watch Abdullah's moves from close by while sharing the glory of liberating Arab and Islamic soil and preventing the creation of a communist-backed Zionist entity on the Egyptian border'.[14]

DAVID AND GOLIATH

One of the most tenacious myths relating to 1948 is that of 'David and Goliath'—that the Arabs were overwhelmingly stronger militarily than the Yishuv. The truth is more complex. It is true that the Jews fought during early 1948 and prepared for the 15 May Arab invasion believing and feeling that the Arab armies, if not the Palestinian irregulars, were stronger than they, were more numerous and professionally trained and led, and were better armed. And it is true that many in the Yishuv were gripped by a sense of foreboding, of vulnerability and of weakness, and that the Haganah / IDF victories

[13] Shlaim, *Collusion, passim.*
[14] Thomas Mayer, 'Egypt's 1948 Invasion of Palestine', *Middle Eastern Studies*, Jan. 1986. The question of the evolution down to 15 May 1948 of the Arab war plans, and their objective/s, has never been properly studied.

in the spring and summer of 1948 were greeted by many as nothing short of miraculous. And, to be sure, in certain engagements—as during the fall of the Etzion Bloc and Kibbutz Degania's repulse of the invading Syrian armoured column, both in May—the Haganah suffered from a significant inferiority in both manpower and weaponry.

But the truth—as conveyed by Milstein, Shlaim, Pappe, and my own work, and several recent Israeli military histories—is that the stronger side, in fact, won. The atlas map showing a minuscule Israel and a giant surrounding Arab sea did not, and, indeed, for the time being, still does not, accurately reflect the true balance of military power in the region. Nor do the comparative population figures; in 1948, the Yishuv numbered some 650,000 souls—as opposed to 1.2 million Palestinian Arabs and some 30 million Arabs in the surrounding states (including Iraq). The pre-May 1948 Yishuv had organized itself for statehood and war; the Palestinian Arabs, who outnumbered the Jews by two to one, had not. The Jews had perhaps not organized optimally, but they had organized. And in war organization, command, and control are almost everything.

During the first half of the war, from December 1947 to 14 May 1948, which was an admixture of a guerrilla and civil war, the Yishuv was better armed, and had more professional officers and better trained 'soldiers' than the Palestinians, whose forces were beefed up by several thousand 'volunteers' (some of them with military training) from the surrounding Arab states. The Haganah's superior organization, command, and control during these first months of war meant that at almost every decisive point of battle, the Jews fielded larger, better-armed, and better-trained formations than their Palestinian antagonists. When matters came to the test, during the Haganah offensives of April and early May, the Palestinian irregulars (and the 'volunteers') collapsed and fled, their redoubts —the Jerusalem Corridor, Tiberias, Haifa, Safad, Eastern Galilee —falling one after another, swiftly and in domino fashion. When one adds to this the Yishuv's superior motivation—it was a bare three years after the Holocaust, and the Haganah troops knew that it was do-or-die—one can see that the Palestinians never had a chance.

The old historiography is no more illuminating when it comes to the second half of the war, the conventional battles between 15 May

1948 and January 1949. Jewish organization, command, and control, reinforced by short, internal lines of communication, were clearly superior to those of the uncoordinated armies of Egypt, Syria, Iraq, and Lebanon; and throughout, the Haganah / IDF also had an edge in numbers. For example, in mid-May, the Haganah fielded some 35,000 armed troops as compared with the 25–30,000 of the Arab invading armies. By the time of Operation Dani, in July, the IDF had 65,000 men under arms and by December, close to 90,000 men under arms—at each stage significantly outnumbering the combined strength of the Arab armies ranged against them in Palestine. As opposed to the Haganah / IDF short lines of communications, the Egyptians and Iraqis had to send supplies and reinforcements over hundreds of kilometres of desert before these reached the front lines.

Two caveats must be entered here. First, Transjordan's Arab Legion was probably the best army in the war, and, man for man, was more effective than most of the Haganah / IDF units it faced. But it numbered only about 5,000 troops, and had no tanks or aircraft (though it had effective, gun-mounted, armoured cars). Second, in terms of equipment, during the crucial three weeks between the pan-Arab invasion of 15 May and the start of the First Truce on 11 June, the Arab armies had a clear edge in weaponry over the Haganah / IDF (the Haganah became the IDF on 31 May). The Haganah was much weaker in the air, had no artillery to speak of (only heavy mortars), and very few tanks and tracked vehicles. For those three weeks, it was 'fifty-fifty', as the Haganah / IDF chief of operations, Yigael Yadin, told the Yishuv's political leadership on 12 May. It was then that superior Jewish motivation, courage, and fortitude made all the difference, carried the day and stopped the invaders almost dead in their tracks. But during both the unconventional war before 15 May, and the renewed hostilities after the First Truce, it was superior Jewish firepower, manpower, organization, and command and control that determined the outcome of battle. (Throughout, the Arab edge in regular artillery was offset by the Yishuv's preponderance in mortars.)

The old histories tend to lay a great deal of emphasis on the Haganah / IDF's perennial shortages of weaponry and ammunition, in large measure due to Western arms embargoes against the Jewish state. The 1948 material from the British and American state archives and the new histories tend to balance this by highlighting the Arab states' grave arms and ammunition purchasing problems during 1948. They, too, suffered severely from intermittent or continuous arms embargoes and politically motivated delays in delivery. For example, one of the reasons the Transjordanians lost or failed to attempt to retake Lydda and Ramle in July was a severe shortage of shells for the Legion's 25-pounders.

It is worth adding to this that the Yishuv, by contrast with both the Palestinian Arabs and the surrounding Arab states, had a relatively advanced, if small, arms-production capacity. For instance, between October 1947 and July 1948, the Haganah's factories produced 3 million 9mm bullets, 150,000 Mills grenades, 16,000 Sten sub-machine-guns, and 210 3-inch mortars.[15]

THE BIRTH OF THE PALESTINIAN REFUGEE PROBLEM

Apart from the birth of the State of Israel, the major political outcome of the 1948 war was the creation of the Palestinian refugee problem. How the problem came about has been the subject of heated controversy between Israeli and Arab propagandists for the past four decades. The controversy has often been heated because it has been perceived as being as much about the nature of Zionism as about what exactly happened in 1948. If the Arab contention is true —that the Yishuv had always intended forcible 'transfer' and that in 1948 it had systematically and forcibly expelled the Arab population from the areas that became the Jewish state—then Zionism is a robber ideology and Israel a robber state. If, on the other hand, one accepts that the refugee exodus was essentially a result of the war, that the war was the handiwork of the Arabs, that the Palestinian masses fled by and large 'voluntarily' or at the behest of their leaders, then Israel emerges free of the taint of what some have called original sin.

In tackling the subject of the exodus historiographically, I can see no escape from recourse to my book, *The Birth of the Palestinian Refugee Problem, 1947–1949*. The only book that previously dealt with the subject in a relatively serious manner was Rony Gabbay's *A Political Study of the Arab–Jewish Conflict: The Arab Refugee Problem (a Case Study)* (Geneva and Paris, 1959). Gabbay's study,

---

[15] Morris, *The Birth*, p. 22.

while pioneering, inevitably suffered from 'prematurity' in the sense that the author had almost no access to contemporary (1948) documentation and perforce had to rely largely on newspaper accounts and a handful of memoirs to describe and explain the exodus. As the subject was highly sensitive, both the newspaper accounts and the recollections of participants, many of them politicians, on whom Gabbay had to rely tended to be polemical and partisan rather than factual, tendentious rather than objective or illuminating. When taking this crucial factor into account, Gabbay's study stands out as a remarkable achievement.

However, the contemporary Israeli, British, and American documentation declassified during the past few years now enables the historian to arrive at a much fuller and more accurate description and explanation of what happened. As I have set out in great detail in *The Birth*, what occurred in 1948 lies somewhere in between the Jewish 'robber state' and the 'Arab orders' explanations. While from the mid-1930s most of the Yishuv's leaders, including Ben-Gurion, wanted to establish a Jewish state without an Arab minority, or with as small an Arab minority as possible, and supported a 'transfer solution' to this minority problem, the Yishuv did not enter the 1948 war with a master plan for expelling the Arabs, nor did its political and military leaders ever adopt such a master plan. What happened was largely haphazard and a result of the war. There were Haganah / IDF expulsions of Arab communities, some of them at the initiative or with the *post facto* approval of the cabinet or the defence minister, and most with General Staff sanction— such as the expulsions from Miska and Ad Dumeira in April; from Zarnuqa, Al Qubeiba, and Huj in May; from Lydda and Ramle in July; and from the Lebanese border area (Kafr Bir'im, Iqrit, Al Mansura, Tarbikha, Suruh, and Nabi Rubin) in early November. But there was no grand design, no blanket policy of expulsion. The best illustration of this, perhaps, is the haphazardness of what happened in the Upper Galilee in late October and early November of 1948.

At the same time, at no point during the war did the Arab leaders issue a blanket call to Palestine's Arabs to leave their homes and villages and wander into exile. Nor was there an Arab radio and press campaign urging or ordering the Palestinians to flee. Indeed, I have found no trace of such a campaign, and had it taken place, had there been such broadcasts, they would have been quoted or at least left traces in the documentation. The Yishuv's intelligence services (the Shai, the Arab Division of the Jewish Agency Political Department, the IDF Intelligence Service, and the Foreign Ministry Middle East Affairs and Research Departments) and the British and American Middle East diplomatic posts all monitored Arab radio broadcasts. So did the BBC. But none of these, in the thousands of monitoring reports, ever refer to, let alone cite from, such an alleged broadcast. Not even once.

Indeed, in early May 1948, when, according to Israeli propaganda and some of the old histories, such a campaign of broadcasts urging or ordering the Arabs to leave should have been at its height, in preparation for the pan-Arab invasion, Arab radio stations and leaders (Radio Ramallah, Radio Damascus, King Abdullah, and Arab Liberation Army chief Fawzi al Qawuqji) all issued calls, in repeated broadcasts, to the Palestinians to stay put or, if already in exile, to return to their homes. Reference to and quotations from these broadcasts appear in a number of Haganah, Mapam, and British documents from early May 1948.

Certainly, local Arab commanders and politicians in a number of areas and on certain occasions ordered the evacuation of women and children from war zones or potential war zones. This happened in areas of East Jerusalem, and in certain villages in the Coastal Plain and in the Galilee in April and May. Less frequently, as happened in Haifa on 22 April, the local Arab leaders advised or instructed their communities to leave rather than stay in a potential war zone or 'treacherously' remain under Jewish rule. But there were no blanket Arab orders or campaigns to leave, neither by the Palestinian national leadership, the Arab Higher Committee, nor by the leaders of the neighbouring Arab states.

Rather, in order to understand the exodus of the 600,000–760,000 Arabs from the areas that became post-1948 Israel, one must look to a variety of related processes and causes.

What happened in Haifa is illustrative of the complexity of the exodus. The evacuation of Haifa's 70,000 Arabs (the city had about an equal number of Jews), as from the other main Palestinian centres, Jaffa and Jerusalem, began in December 1947 with the start of sporadic hostilities between neighbouring Jewish and Arab districts. The exodus slowly gained momentum during the following months

as the British Mandate administration began to wind down and
prepare for withdrawal from Palestine. The first to leave were the
rich and the educated—the middle and upper classes with money to
stay in hotels or with second homes on the Beirut seafront, in
Nablus and in Amman, or with comfortably-off relatives abroad
with large homes. The Palestinians' political and economic leadership
disappeared. By mid-April, the bulk of the members of the National
Committees of Haifa and Jaffa had left the country. By mid-May,
only one member of the Arab Higher Committee (Dr Husayn
Fakhri al Khalidi, of Jerusalem) was still in the country.

The gradual flight of the professionals, the civil servants, the
businessmen, the traders and the shopkeepers, the doctors and
lawyers and chemists, over December 1947–April 1948 had a
cumulative, harsh impact on the Haifa Arab masses, who were at
the same time demoralized by the incessant sniping and bomb
attacks, by the interdiction of supply and reinforcement convoys, by
price rises and unemployment, by the feeling that the Jews were
stronger (and held the dominant, high ground), and by the sense
that their own ragtag militia would fail when the test came (as,
indeed, it did). The Arabs felt isolated and insecure, at the mercy of
their neighbours up the Carmel and of the kibbutzim lying astride
the city's approach roads. The impending British evacuation and
the dissolution of British-led government bred an all-pervading fear
of the future. Businesses, shops, and workshops closed, policemen
shed their uniforms and left their posts, Arab workers could no longer
commute to jobs in the Jewish districts, and agricultural produce
was occasionally interdicted in ambushes on the approach roads.
There was increasing unemployment and there were sharp price-
increases. By mid-April, some 15,000–20,000 Arabs, apparently,
had left the city.

Then came the Haganah assault on the Arab districts of 21–2
April. Several companies of Carmeli Brigade troops (supported by
a company of Palmahniks), under cover of a constant mortar
barrage, drove down the Carmel Mountain slopes into the Arab
downtown areas. Arab militia resistance collapsed. Thousands
of Arabs fled from the attacked outlying neighbourhoods (Wadi
Rushmiya, Halissa) into the British-controlled port area, piled into
boats, and fled northwards, to Acre. In their passage through the
central Arab areas—Wadi Nisnas, Wadi Salib, the Suq—the refugees

from the outlying neighbourhoods spread 'flight fever' among the downtown-district Arabs. The local leaders who remained sued for a ceasefire. Under British mediation, the Haganah agreed, offering what the British regarded as generous terms. But then, when faced with Moslem pressure, the largely Christian leadership got cold feet; a ceasefire meant surrender and implied readiness to live under Jewish rule. They would be open to charges of collaboration and treachery. So, to the astonishment of the British and the Jewish military and political leaders gathered on the afternoon of 22 April at the Haifa town hall, the Arab delegation announced that its community would evacuate the city.

The Jewish mayor, Shabtai Levy, and the British commander, Major-General Hugh Stockwell, pleaded with the Arabs to reconsider; the joint Jewish–Arab municipality had functioned amicably and efficiently for decades. The Haganah representative, Mordechai Makleff (whose parents had been butchered by Arab rioters in 1929), declined to voice an opinion. But the Arabs were unmoved, and the mass exodus, which had begun already during the morning under the impact of the Haganah mortars and ground assault, moved into top gear, with the British supplying boats (Z-craft) and an armoured-car escort to the departing Arab convoys. From 22 April to 1 May, almost all the Arab population departed the city. The rough treatment—temporary evictions, house-to-house searches, detentions of able-bodied males, and occasional beatings—meted out to the remaining population during that week by the Haganah and the Irgun Zva'i Leumi (IZL) troops who occupied the downtown areas led many of the undecided also to opt for evacuation. By early May, the city's Arab population had dwindled to three or four thousand.

In some respects, the experience of Arab Haifa was unrepresentative: Nowhere else in the country in April, or later, did Jewish leaders plead with their Arab neighbours to stay put; only in a few other places (Tiberias, Qumiya) did the British military assist the Arabs to evacuate; in few places was there an organized, considered decision by the local leadership to evacuate. But in most major respects, Haifa was typical: the middle-class flight months before the assault; the growing, general sense of insecurity; the expanding unemployment and the rising food prices; the cowardly disappearance before and during the battle of the local leaders; the mass flight under the impact and largely as a result of the Haganah assault.

The bulk of the Palestinian refugees—some 250,000–300,000—went into exile during those weeks between April and mid-June 1948, with the major precipitant being (Haganah / IDF or IZL) military attack or fears of such attack. In most cases, the Jewish commanders, who preferred to occupy empty villages (occupying populated villages meant necessarily leaving behind a garrison, which the units could ill afford to do), were hardly ever confronted with the decision whether or not to expel an overrun community: most villages and towns simply emptied at the first whiff of grapeshot.

In conformity with *Tochnit Dalet* (Plan D), the Haganah's master plan, formulated in March 1948, for securing the Jewish state areas in preparation for the expected declaration of statehood and the prospective Arab invasion, the Haganah cleared various areas completely of Arab villages—the Jerusalem corridor, the area around Mishmar Ha'emek, and the coastal plain. But in most cases, expulsion orders were unnecessary; the inhabitants had already fled, out of fear or as a result of Jewish attack. In several areas, Israeli commanders successfully used psychological warfare ploys to obtain Arab evacuation (as in the Hula Valley, in Upper Galilee, in May).

The basic structural weaknesses of Palestinian society led to the dissolution of the society when the test of battle came. Lack of administrative structures for self-rule, as well as weak leaders, poor or non-existent military organization beyond the single-village level, and faulty or non-existent taxation mechanisms, all caused the main towns to fall apart in April–May, as the British withdrew and the Haganah attacked. The fall of the towns and the exodus from them, in turn, caused fear and despondency in the rural hinterlands. Traditionally, the villages, though economically autarchic, had looked to the towns for political leadership and guidance. The evacuation by the urban middle classes and leaders (who often owned lands in the countryside and had rural origins), as well as the fall of the towns, provided the Palestinian villagers with an example to emulate. Safad's fall and evacuation on 10–11 May, for example, triggered an immediate evacuation of the surrounding Arab villages; so, earlier, did the fall of Haifa and the IZL assault on Jaffa.

Seen from the Jewish side, the spectacle of mass Arab evacuation certainly whetted appetites for more of the same. Everyone, at every level of military and political decision-making, understood that a Jewish state without a large Arab minority would be stronger and more viable both militarily and politically. The tendency of local military commanders to 'nudge' Palestinians into flight increased as the war went on. Jewish atrocities—far more widespread than the old histories have let on (there were massacres of Arabs at Ad Dawayima, Eilaboun, Jish, Safsaf, Majd al Kurum, Hule (in Lebanon), Saliha, and Sasa, besides Deir Yassin and Lydda and other places)—also contributed significantly to the exodus.

FAILED PEACE EFFORTS

The last major myth tackled incidentally or directly by the new historians concerns an Israel that in 1948 and 1949 was bent on making peace with its neighbours, and an Arab world that mono-lithically rejected all such peace efforts.

The evidence that Israel's leaders were not desperate to make peace and were unwilling to make the large concessions necessary to give peace a chance is overwhelming. In Tel Aviv there was a sense of triumph and drunkenness with victory—a feeling that the Arabs would 'soon' or 'eventually' sue for peace, that there was no need to rush things or offer concessions, that ultimately military victory and dominance would translate into diplomatic-political success (peace).

As Ben-Gurion told an American journalist in mid-July, 1949: 'I am prepared to get up in the middle of the night in order to sign a peace agreement—but I am not in a hurry and I can wait ten years. We are under no pressure whatsoever.'[16] Or, as Ben-Gurion records Abba Eban, Israel's ambassador to the UN, saying: '[Eban] sees no need to run after peace. The armistice is sufficient for us; if we run after peace, the Arabs will demand a price of us—borders [that is, in terms of territory] or refugees [that is, repatriation] or both. Let us wait a few years.'[17]

---

[16] Quoted in Shlaim, *Collusion*, p. 465.
[17] David Ben-Gurion, *Yoman Hamilhama-Tashah* [the war diary 1948–9], ed. Gershon Rivlin and Elhannan Orren (Tel Aviv, 1982) iii, 993.

As Pappe puts it:

Abdullah's eagerness [to make peace] was not reciprocated by the Israelis. The priorities of the state of Israel had changed during 1949. The armistice agreements brought relative calm to the borders, and peace was no longer the first priority. The government was preoccupied with absorbing new immigrants and overcoming economic difficulties.[18]

Israel's tergiversation or lack of urgency regarding peace was manifested most clearly in the protracted (1948–51) secret negotiations with Abdullah. Israeli Foreign Minister Moshe Sharett thus described his meeting with Transjordan's king at the palace in Shuneh on 5 May 1949: 'Transjordan said—we are ready for peace immediately. We said—certainly, we too want peace but one shouldn't rush, one should walk.'[19]

Israel and Transjordan eventually signed an armistice agreement, on 3 April 1949. But it had involved a great deal of brow-beating and arm-twisting by Israel (which, with unmerited exaggeration, British and American observers likened to Hitler's treatment of the Czechs in 1938–9). Abdullah took it realistically: 'If you meet a bear when crossing a rotten bridge, call her "Dear Auntie"', he said, quoting a Turkish adage. But the two countries never signed a peace treaty or a non-belligerency pact, as was suggested by Abdullah at one point.

Shlaim, who in *Collusion* considerably expands the description of the secret Israeli–Transjordanian negotiations first covered in Dan Shueftan's *Ha'Optziya Ha'Yardenit* (The Jordanian Option), published in 1986, more or less lays the blame for the failed talks on Israel's shoulders. At one point, it is true, Shlaim writes: 'Two principal factors were responsible for the failure of the post-war negotiations: Israel's strength and Abdullah's weakness.'[20] But more often, Shlaim attributes more weight to the former and less to the latter, and the reader comes away with the clear impression of overwhelming Israeli responsibility for the failure. Israel refused to offer major concessions in terms of refugee repatriation or territory (Abdullah was particularly keen on getting back Lydda and Ramle)

[18] Pappe, *Britain and the Arab–Israeli Conflict*, 188.
[19] ISA, *Documents on the Foreign Policy of Israel, May–December 1949*, iv. 68, 'M. Sharett's Address at a Meeting of the Heads of Divisions in the Ministry of Foreign Affairs (Tel Aviv, 25 May 1949)'.
[20] Shlaim, *Collusion*, p. 621.

and was for too long unwilling to offer Jordan a sovereign corridor through its territory to the sea at Gaza (or Ashkelon). Throughout, says Shlaim, Israel was prodded if not guided by the 'blatant expansionism' of some of Ben-Gurion's aides, such as Moshe Dayan. As one of Dayan's colleagues, Yehoshafat Harkabi, later put it: 'The existential mission of the State of Israel led us to be demanding and acquisitive, and mindful of the value of every square metre of land.'[21] Ben-Gurion declined to meet with Abdullah and often the Hashemite monarch was spoken of in Tel Aviv with undeserved contempt.

Yet while all this was true, Shlaim fails sufficiently to acknowledge the importance of the 'Palestinization' of Jordan following the Hashemite annexation of the West Bank, which quickly resulted in the curtailment of Abdullah's decision-making autonomy and of his freedom of political movement both within Jordan and in the pan-Arab arena. The twin pressures exercised by his successive cabinets in Amman and by the Arab world outside successfully impeded Abdullah's ability to make a separate peace with Israel. He almost did so a number of times, but always he held back at the last moment and declined to take the plunge. It is possible, argues Shlaim, that more generous, up-front concessions by Tel Aviv at certain critical points in the negotiations would have given Abdullah greater motivation to pursue peace as well as the ammunition he needed to silence his anti-peace critics, but one will never know. What is certain is that Abdullah, though showing remarkable courage throughout, simply proved unable during his last years to go against the unanimous or near-unanimous wishes of his ministers and against the unanimous or near-unanimous anti-peace stance of the surrounding Arab world.

What happened with Abdullah occurred in miniature and more briefly with Egypt and Syria. During September–October 1948, Egypt's King Farouq, knowing that the war was lost, secretly sent a senior court official to Paris to sound out Israel on the possibility of reaching a peace based on Israeli cession of all or parts of the Negev and of recognition of Egyptian annexation of the Gaza Strip. Sharett and his senior Foreign Ministry advisers favoured continued negotiations, but Ben-Gurion flatly rejected the overtures. The

---

[21] Shlaim, *Collusion*, pp. 444–5.

Cabinet on 6 October had voted for a renewed push against the Egyptians in the Negev and Ben-Gurion did not want possibly inconsequential peace talks to put a spoke in the wheels of the planned military offensive (Operation Yoav, which was launched on 15 October, and which resulted in the IDF punching a broad hole through the Egyptian-held Gaza–Faluja–Hebron line and relieving the Egyptian siege of Israel's two dozen Negev kibbutzim). It is possible that Farouq had ordered the peace feelers specifically in order to prevent Israel from launching the offensive rather than out of any real desire or intention to make a separate peace with the Jewish state.

Shlaim summarizes the episode thus: '[Ben-Gurion] may have been right in thinking that nothing of substance would come out of these talks. But he surely owed his cabinet colleagues at least a report on what had taken place so that they could review their decision [of 6 October] to go to war against Egypt on the basis of all the relevant information.'[22] New Egyptian peace feelers in November, in the middle of Operation Yoav, again came to nought as a result of Ben-Gurion's veto.

As for Syria, in May 1949 its new ruler, Colonel Husni Za'im, made major peace proposals which included recognition of Israel as well as Syrian readiness to absorb hundreds of thousands of Palestinian refugees. Za'im wanted Israel to concede a sliver of territory along the Jordan River (and perhaps half the Sea of Galilee). He asked to meet with Ben-Gurion. Ben-Gurion rejected the proposal, writing on 12 May:

'I am quite prepared to meet Colonel Za'im in order to promote peace . . . But I see no purpose in any such meeting so long as the representatives of Syria in the armistice negotiations do not declare in an unequivocal manner that their forces are prepared to withdraw to their pre-war territory' (that is, withdraw from the small Syrian-occupied Mishmar Ha'Yarden salient, west of the Jordan).[23]

Continued feelers, via the UN, by Za'im resulted again in Israeli stonewalling. As Sharett put it on 25 May: 'It is clear that we . . . won't agree that any bit of the Land of Israel be transferred to Syria,

---

[22] Shlaim, *Collusion*, p. 320.
[23] ISA, *Documents on the Foreign Policy of Israel*, iii. 563, R. Shiloah to R. Bunche, 12 May 1949.

because this is a question of control over the water sources [of the Jordan River].'[24] Za'im declined to meet Sharett and reiterated his wish to meet 'only [the] Prime Minister'.[25] Shabtai Rosenne, the legal adviser at the Israel Foreign Ministry, put it simply: 'I feel that the need for an agreement between Israel and Syria pressed more heavily on the Syrians.'[26] Therefore, why rush towards peace? A few weeks later Za'im was overthrown and executed; the Syrian peace initiative died with him. Whether the overture was serious or merely tactical—to obtain Israeli territorial concessions in the armistice agreement, or to obtain Western support and funds for his shaky regime—is unclear. What is certain is that Israel failed wholeheartedly or even half-heartedly to pursue it.

What was true of Israel's bilateral contacts with each of the main Arab states was true also of its negotiations with the Arab states under UN auspices at Lausanne in the spring and summer of 1949. There, the Arab states all too often exhibited obdurate, unconciliatory, and unrealistic positions, lining up, in a show of singular behaviour, behind the most extreme of their number. The Arabs spoke of a repatriation of all the refugees and a return to the 1947 partition borders, or worse; a few delegates—the Transjordanians, for example—in private spoke of a return of 'only' 400,000 refugees.

But Israel, too, proved ungenerous and unconciliatory. For months, UN and US officials pressed Tel Aviv to make a 'saving' gesture, a redemptive concession: to proclaim its willingness to take back 200,000–250,000 refugees; the Arabs would then be persuaded to take and resettle the rest. The months dragged on and neither side budged. When, under intense US–UN prodding, Israel at last offered to take back '100,000' (which, in reality was only 65,000, as Sharett explained to his colleagues), it was a clear case of 'too little, too late'. And Israel's more realistic offer, to accept the Gaza Strip

[24] ISA, *Documents on the Foreign Policy of Israel*, iv. 69, 'M. Sharett's Address at a Meeting of the Heads of Divisions in the Ministry of Foreign Affairs (Tel Aviv, 25 May 1949).

[25] ISA, *Documents on the Foreign Policy of Israel*, iii. 592, Shiloah to Sharett, 31 May 1949.

[26] ISA, *Documents on the Foreign Policy of Israel*, iii. 582, S. Rosenne to D. Ben-Gurion and M. Sharett, 18 May 1949. The most comprehensive review of the evidence concerning the Za'im peace initiative of 1949 is to be found in Arye Shalev, *Shituf Pe'ula Betzel Imut* [co-operation under the shadow of conflict] (Tel Aviv, 1989), 49–65. Shalev, like Shlaim, is critical of Israel's lack of eagerness in responding to the Za'im initiative.

along with its 200,000-odd refugees, was never seriously entertained by Egypt. So the prospect of a comprehensive Israeli–Arab peace was buried at Lausanne for at least forty years.[27]

There are people who use history to prove political points and buttress contemporary political positions; there are others who would brandish history in the service of present-day peacemaking. For myself, I am not sure that writing history serves any purpose or should serve any purpose that strays beyond the covers of each book and beyond the desire of practitioners and readers to penetrate the murk of the past. I am fairly confident that the new Israeli historiography—and I am not sure that it has a Palestinian counterpart—has succeeded, at least to some degree, in penetrating that murk and illuminating it to a degree never before attained, and that a more accurate and realistic understanding of the roots of the Israeli–Arab conflict is now emerging.

Addendum (October 1989)

The publication of my *Birth of the Palestinian Refugee Problem, 1947–1949*, of Shlaim's *Collusion Across the Jordan*, and of the original version of this article on the New Historiography in the American Jewish bi-monthly *Tikkun* (issue of November–December 1988) triggered a major controversy in Israeli and Diaspora-Jewish academic circles. The New Historiography was roundly discussed in a two-day conference at Tel Aviv University in April 1989, and was comprehensively attacked in a series of articles (about 20,000 words *in toto*) by Ben-Gurion's biographer, Shabtai Teveth, in the Israeli daily *Ha'aretz* on 7, 14, and 21 April and 19 May 1989. My reply appeared in *Ha'aretz* on 9 May, with a short addendum in early June. Teveth subsequently published an abridged, amended version of the series in the American monthly, *Commentary*, in September 1989.

Several of the speakers at the TAU conference said that the historiography in question could not properly be referred to as New History (comparable, say, to the New History recently in vogue in France) as it did not offer a new outlook on history or new tools or areas of historical research but simply constituted a new interpretation within the bounds of traditional historiography. One generally sympathetic critic, Professor Yehoshua Porath of the Hebrew University, suggested that instead of contrasting New Historiography with Old Historiography, it would be more accurate to speak of

[27] See Morris, *The Birth*, chap. 9, *passim*.

plain History versus Pre-History: what is currently being written in Israel is History, said Porath, whereas in the 1950s, 1960s, and 1970s what was written, by and large, was Pre-History. I have no objection to Porath's suggested definitions or re-definitions.

Professor Itamar Rabinovich, of Tel Aviv University's Dayan Centre, argued that peace agreements with the Arab states were not reached in the immediate wake of the 1948 war not because of Israeli tergiversation or intractability but essentially because of Arab 'rejectionism' and intransigence. Rabinovich attacked Shlaim's claim that it was Israeli obduracy that stymied the peace process with Abdullah during 1949–51 and that it was Israeli indifference that frustrated Syrian President Za'im's pacific overtures in mid-1949. Za'im, who only ruled Syria for four months, was a lightweight who could not have carried through a peace process even had he been serious—and his overtures do not appear to have been serious, argued Rabinovich.

Shlaim responded that Israel had not been conciliatory enough or quick enough with its concessions in the negotiations with Jordan and that this was a key element in the failure of the talks before Abdullah's assassination. As to Za'im, argued Shlaim, while it is not clear whether Za'im's overtures had been serious, Israel should have vigorously pursued them if only to test the Syrian leader's sincerity. Ben-Gurion and Sharett failed to do this, as they had failed to pursue earlier peace feelers put out by Egypt in September and November 1948.

To turn to Teveth's articles. Teveth attacks the New Historiography on both political and historiographic grounds. He maintains that one of the characteristics of the 'new historical club . . . is a sympathy somewhat inclined to the side of the Palestinians'—or, to put it more simply, the new historians are pro-Palestinian. Moreover, says Teveth, their aim is to 'deligitimise Zionism'. To demonstrate this, Teveth quotes from my original *Tikkun* essay a passage stating that 'how one perceives 1948 bears heavily on how one perceives the whole Zionist / Israeli experience . . . If Israel was born tarnished, besmirched by original sin, then it was no more deserving' of Western sympathy and aid than its Arab neighbours. Teveth goes on to say that the 'original sin' Shlaim and Morris charge Israel with consists of 'the denial to the Palestinian Arabs of a country' or state, and 'creating the refugee problem'. Both charges, says Teveth, 'are false'.

Later in the article Teveth quotes from a favourable reference to my book by Professor Edward Said, the Columbia University English literature teacher, PLO propagandist, and member of the Palestine National Council. Damning by association. The new history, concludes Teveth, is bent on 'providing fresh sources of political sympathy for the Arabs and fresh sources of antipathy to the Jews'. Ultimately, what Teveth is saying is that the new Israeli historiography is unpatriotic or anti-patriotic and is serving Arab or Palestinian Arab political ends.

My response to this, in *Ha'aretz* and in a second article in *Tikkun*, published in January–February 1990, was that the New Historiography, as far as I was concerned, had no political purposes whatsoever. The task and function of the historian, in my view, is to illuminate the past—to describe what happened and explain why things happened as they did, taking account of the motives and considerations of the various protagonists. In *The Birth of the Palestinian Refugee Problem, 1947–1949* I tried to describe and explain how the refugee problem was created. I evinced no sympathy for either side—indeed, the portrayal of the Palestinians, many Palestinians of my acquaintance have told me, is far from flattering—though I tried to explain why each side acted as it did during the prolonged process. If anything, most of my readers probably came away with the feeling that both Jews and Palestinian Arabs acted in an almost inevitable, understandable manner and, in the circumstances of 1948, could probably have acted no differently. I did not try or aim to delegitimize Zionism, which I regard as legitimate a national movement as the next one. Perhaps a world without nationalism would be a pleasanter place to live in; or perhaps not. In any event, Zionism's claim to legitimacy is as cogent as that of most national movements and, given the background of the Holocaust and the circumstances of 1948, was certainly more pressing and immediate than most.

In *The Birth* I rendered no moral or value judgements. I used the terms 'pure', 'innocent', and 'original sin' in the *Tikkun* article to explain not history or my own views but how others—specifically Israeli and Arab propagandists, who have over the decades viewed the conflict in black and white—have regarded the Arab–Israeli conflict and 1948. *The Birth*, one might say, occupies the grey, rather than the moral, high ground.

The possibility that his findings and conclusions might subsequently be used by propagandists and politicians of this or that ilk is surely no concern of the historian. If he begins to 'streamline' his descriptions, analyses, or conclusions according to this or that contemporary political calculation, sensibility, or interest, he will produce less, perhaps far less, than unbiased, impartial history. The historian must be judged according to the degree of accuracy and understanding he has achieved regarding the past—nothing else.

Teveth's 'political' objections and views seem to underlie his 'historiographic / professional' critique of the New Historiography in general and *The Birth* in particular.

The focus of Teveth's criticism of *The Birth* is on the causation of the Arab exodus from Palestine up until 15 May 1948. He writes in *Commentary*: 'before the invasion [of Palestine by the Arab armies on 15 May] the

Palestinian Arabs were seen [by the Yishuv], for the most part, as citizens of a future Jewish state; after it, as declared enemies. Accordingly one may properly speak, in the former period, of an Arab flight, and in the latter of expulsion by Israel.'

To explain the flight up to 15 May Teveth quotes from a speech by Ben-Gurion in the Knesset on 11 October 1961:

The Arabs' exit from Palestine . . . began immediately after the UN [partition] resolution . . . And we have explicit documents testifying that they left Palestine following instructions by the Arab leaders, with the Mufti at their head, under the assumption that the invasion of the Arab armies . . . would destroy the Jewish state and push all the Jews into the sea . . .

Teveth goes on to cite from the single piece of evidence he found—which, incidentally, he calls 'one [document] in particular', hinting that other, similar ones exist—that supports his (and Ben-Gurion's avowed) view that the Arabs of Palestine who fled their home, villages, and towns before 15 May (about half the total that left the country during the war) did so on instructions from their leaders. The document, the 'Haganah Intelligence Service Daily Report' of 24 April 1948, states: 'Rumours have it that the Arab Higher Committee in Jerusalem ordered the evacuation of Arabs from several localities in Palestine . . . Arab residents are advised to flee Palestine as soon as possible, and after its fall into the hands of the Arab governments, they will be returned as victors.'

Earlier, Teveth quotes from other Haganah intelligence reports that refer to specific orders by Arab leaders and commanders to a handful of villages to evacuate and to other orders to several local councils to evacuate women and children from combat zones or potential combat zones. (These documents are all cited and quoted from in *The Birth*.)

All this evidence, taken together, shows, according to Teveth (in *Commentary*) that 'the Arab exodus from Palestine . . . was all the work of instruction, whether by personal example, by word of mouth, or in writing, or even better, by the quickest telegraph of all, rumour.' (This formulation first appeared in Teveth's article in *Ha'aretz* of 14 April. Interestingly, after I had taken him to task for its relative obscurity and imprecision (which enabled Teveth to avow the 'Arab orders' explanation for the exodus without saying so clearly), Teveth abridged it somewhat, writing in *Ha'aretz* on 19 May that the Arab flight before 15 May had been precipitated by 'an order by personal example' (implying, apparently, that the Arab flight had been triggered by the exemplary flight of their local leaders and that this was tantamount to an order to flee by the Arab leadership).

In my rebuttal (in *Ha'aretz*, 9 May 1989, and in *Tikkun*, January–February 1990) of Teveth's explanation of the first half of the Arab exodus,

I argued or, rather, reiterated that the exodus had been a complex, multi-staged phenomenon which had included, among its causes, the innate weaknesses of Palestinian Arab society (poor leadership and a lack of governmental structures, poor military organization and lack of self-confidence); material privations (rocketing prices and mass unemployment), in part stemming from the widening circle of hostilities; the gradual devolution of British rule and the spreading breakdown of law and order, especially in the towns and cities; a general fear of the future and a specific fear of Jewish conquest and rule; the earlier, demoralizing departure of the Arab middle class and the élite, including, in many cases, their local political and military leaders; actual and feared atrocities; Arab orders to specific communities to evacuate; specific Jewish expulsion orders; and Haganah / IDF assault and conquest or fears of imminent Jewish assault. I argued that this final cause—Jewish assault and conquest—was in most cases the main and final precipitant to flight.

In refuting Teveth's single-cause ('Arab orders') explanation of the exodus up to 15 May, I pointed out that there is simply no evidence to support it, and that the single document Teveth is able to cite, the Haganah report of 24 April, refers explicitly to 'rumours' and to an order to 'several localities' (rather than a blanket order to 'the Arabs of Palestine'). Moreover, neither these 'rumours' nor the purported order were referred to again in any subsequent Haganah intelligence reports (which surely would have been the case had these 'rumours' been confirmed and had an actual order been picked up). The fact is that the opposite occurred: Haganah intelligence and Western diplomatic missions in the Middle East at this time, around 5–6 May 1948, picked up, recorded, and quoted from Arab orders and appeals (by King Abdullah, Arab Liberation Army Commander Fawzi Qawuqji, and Damascus Radio) to the Arabs of Palestine to stay put in their homes or, if already in exile, to return to Palestine. Not evidence of 'Arab orders' to flee but of Arab orders to stay put during those crucial pre-invasion weeks are what one finds in Israeli and Western archives.

But Teveth's thesis is defective not merely because of a lack of evidence. It flies in the face of the chronology, which there is no getting around. There was an almost universal one-to-one correspondence between Jewish attacks in specific localities and on specific towns and Arab flight from these localities and towns. Tiberias was attacked by the Golani Brigade on 17 April; its Arab population evacuated on 18 April. Arab Haifa was attacked and defeated on 21–2 April; most of its 70,000 Arab inhabitants, evacuated the city over 22 April–1 May. Jaffa was assaulted by the Irgun Zva'i Leumi on 25–7 April; the bulk of its 70,000–80,000 population fled the city between 25 April and 13 May. Safad was attacked and conquered by the Palmah on 9–10 May; its Arab population of 10,000 fled the city on 10 May. Eastern

Galilee was conquered by Palmah units between 2 May and 25 May; the villages in the area decamped durirg that period. And so on.

What this means is that Haganah / Irgun / IDF attack was usually the principal and final precipitant of Arab flight. Had, as Teveth argues, a general Arab order or series of orders been the cause and precipitant of flight, the flight would necessarily have occurred hard upon the issuance and publication of that order or series of orders. But the staggered, highly localized nature of the Arab flight from each area wreaks havoc with any such thesis. For if the Arab order / orders had been issued on 10 April, why did the inhabitants of Haifa wait a fortnight, and those of Safad or Eastern Galilee a month and more to depart? And if the order was issued, say, on 25 April, why did the inhabitants of Tiberias depart a week earlier, and those of Haifa begin to depart three days before; or those of Safad wait a further fortnight before leaving? And if the order was only issued on 5 May, why did the inhabitants of Tiberias, Jaffa, Haifa, and the Jerusalem Corridor depart or begin to depart weeks and days before and those of Safad only a week later? And so on.

Teveth could, theoretically, argue that the Arab order or orders to evacuate were issued at or around a specific date, but the Arab inhabitants in some localities pre-empted the order, and in others ignored the order and delayed their departure until the Jews attacked. But this would mean, simply, that the Arab flight was not caused or precipitated by the order / orders.

But the fact is that there was no order / series of orders and there was no co-ordinated, unified exodus from Palestine around a specific date: there was a cumulative process, there were interlocking causes, and there was a main precipitant, a *coup de grâce*, in the form of Haganah / Irgun / IDF assault in each locality.

Teveth, for some reason, dislikes my division of the exodus into four stages—Stage I, December 1947–March 1948; Stage II, April–June 1948; Stage III, July 1948; and Stage IV, October–November 1948. Unfortunately, the flight occurred in these stages, which were defined and circumscribed by the various stages of the war and, from April 1948, by the successive Haganah / IDF offensives, with periods of truce in between. Teveth also charges that I did not divine 'patterns' in the Arab flight—which is nonsense. Hundreds of pages of *The Birth* describe these patterns in great detail (the 'domino' phenomenon, of flight from one village precipitating flight from a neighbouring village; flight from central towns and cities provoking flight from satellite villages; the flight-inducing effects of atrocities; and so on).

Teveth in his articles also attacked my portrait of Ben-Gurion, as he emerges from the pages of *The Birth*. He would have the Yishuv's leader a benign father-figure, slow to anger, who was only propelled into animosity towards the Palestinian Arabs by the 15 May pan-Arab invasion. The truth is somewhat different. Early on Ben-Gurion understood that the prospective Arab minority represented the major, existential threat to a future Jewish state; and he understood, from at least the mid-1930s (as traced in Teveth's own *Ben-Gurion and the Palestinian Arabs*), that the Palestinians were implacably opposed to the emergence of a Jewish state in Palestine and would resist it to the hilt. So, already from 1937 we find Ben-Gurion (and most of the other Zionist leaders) supporting a 'transfer' solution to the 'Arab problem'. (This too is traced in Teveth's *Ben-Gurion and the Palestinian Arabs*, as well as in *The Birth*, pp. 23–8).

Come 1948, and the confusions and displacement of war, we see Ben-Gurion quickly grasp the opportunity for 'Judaizing' the emergent Jewish state. Certainly, he understood the requirements of statesmanship—continued lip service to the enlightened ideas of Western Democracy and Socialist Humanism and the necessity to hide traces of behaviour that others might construe as immoral or hard-hearted. But he also understood that one could not make an omelette without breaking eggs, Arab eggs. So he nudged his officers and officials in the right direction and, in the course of battle, most of the country emptied of its original Arab inhabitants.

Ben-Gurion was certainly a real-politician—devious, sly, and wise, resolute, single-minded, and ruthless: or what, in fact, history requires of successful nation-builders. No responsible Yishuv leader, in the dire straits of post-Holocaust 1948, would or could have acted otherwise. Teveth would have us believe in Ben-Gurion the lily-livered, soft-hearted liberal. But this wasn't Ben-Gurion; certainly not Ben-Gurion in 1948.

Teveth, in *Ha'aretz* and *Commentary*, charges that I portrayed Israel as 'a military "superpower", a juggernaut crushing the Palestinians and the Arab armies alike'—thus demolishing the myth of (the Jewish) David and (the Arab) Goliath. In *Commentary* he details the relative strength of the Arabs in artillery and the relative weakness in artillery of the Haganah / IDF in May–July 1948 in order to demonstrate the continued cogency of the (Jewish) David–(Arab) Goliath view of 1948.

I stand by my original thesis (to be found in *The Birth passim* and in my original article in *Tikkun*): that the Yishuv was vastly superior to the Palestinian Arabs in military strength, and that this superiority—in mobilization of manpower, command and control, numbers and training, and yes, most categories of weapons, together with the ability to concentrate units and weapons (partly due to short lines of communication and to superior organization)—enabled the Haganah, when it went over to the offensive, to crush the Palestinian militias in a relatively short—and bloodless—series of offensives over April–early May 1948. And the

Haganah / IDF was also, for much the same reasons, stronger than the combined Arab expeditionary forces that invaded and fought in Palestine. This superiority, again, was determined by superior command and control; better trained and greater manpower (in mid-May the Haganah fielded some 35,000 troops as opposed to the invading Arab armies' total of 25,000–30,000; in mid-July, the IDF had 65,000 troops; in November, some 90,000, or twice as many as the combined Arab armies at the time); short lines of communication (as opposed to very long ones for the Egyptian and Iraqi expeditionary forces); and a superiority in most categories of weaponry (except during the three critical weeks between 15 May and 11 June). By the offensives of October-November, the IDF's manpower and weapons superiority (which was augmented by the *de jure* or *de facto* British and American boycotts of weapons and ammunition to the Arabs) over the Arab armies was stark and decisive.

Wars are almost invariably won by the stronger side, and 1948 was no exception. This is not to belie the courage and self-sacrifice of the Jewish troops. Courage and self-sacrifice, indeed, were crucial during the three weeks in mid-1948 in which the invading Arab armies were stopped in their tracks; and crucial in the battles of many kibbutzim, such as Yad Mordechai and Negba, in repelling localized attacks. In all, the Yishuv suffered 6,000 dead or one person in every hundred. But the key to the victory of 1948 lay, as always, in the traditional indices of strength—command and control, manpower, and weaponry.

Lastly, Teveth charges that in *The Birth* I exaggerated the importance and influence of Yosef Weitz, the director of the Jewish National Fund's Lands Department, the chairman of the Negev Committee, and the head of the first and second Transfer Committees. I refer the reader to Essay 4 in the present volume, 'Yosef Weitz and the Transfer Committees, 1948–1949'. The reader can then make up his own mind whether the first Transfer Committee in fact existed (Teveth says it didn't) and what, if any, was Weitz's contribution to the transfer of Palestine's Arabs.

2

Mapai, Mapam, and the Arab Problem in 1948

The first Israeli–Arab war, in 1948, resulted in the creation of the Palestinian refugee problem. In great measure, that problem was a product of Jewish military and political actions. To facilitate understanding of these actions, and the attitudes and policies that underlay them, it is worth examining the evolution of attitudes to 'the Arab problem' and to the emerging refugee problem in the political parties of the Yishuv in the course of 1948.

The focus of such an examination must inevitably be on the only two parties that counted for much in the decision-making processes during that war—Mapai (*mifleget poalei eretz yisrael*, the Land of Israel Workers Party) and Mapam (*mifleget poalim meuhedet*, the United Workers Party).

The right-wing revisionists, perhaps representing 10–15 per cent of the electorate, were outside the decision-making process and were only of marginal military importance through their IZL formations. The General Zionists and Progressives, two liberal parties, though represented in the Cabinet, were small and ineffectual. The religious parties, also representing 10–15 per cent of the voters, were ideologically and institutionally completely removed from the mainstream of Yishuv life, especially in wartime.

The 1948 Cabinet, or Provisional Government of Israel as it was called then, had thirteen members. Four of these, holding, among others, the key posts of the Premiership and Defence Ministry (David Ben-Gurion), the Foreign Ministry (Moshe Shertok, later, Sharett) and the Finance Ministry (Eliezer Kaplan), were Mapai members. Two others—Agriculture Minister Aharon Zisling and Labour Minister Mordechai Bentov—were members of the sister socialist party, Mapam. Two other ministers, Police and Minority Affairs Minister Bechor Shitrit and Interior Minister Yitzhak Gruenbaum, who represented respectively the Sephardi (oriental Jewish) community, and a stream of dissident General Zionists, were for all intents and purposes allies or satellites of Mapai and almost invariably sided with the two socialist parties on major issues (when the vote was along party lines). These eight ministers together formed the core of the Provisional Government. The other members of the Cabinet were Pinhas Rosen (Felix Rosenbluett), the Progressive Party's Justice Minister; Peretz (Fritz) Bernstein, the General Zionists' Minister of Industry and Commerce; Yitzhak Meir Levine, the Social Affairs Minister from the ultra-orthodox Agudat Yisrael Party; Moshe Shapira, the Minister of Immigration and Health, from the Mapai-affiliated orthodox Hapoel Hamizrahi Party; and Yehuda Leib Maimon (Fischman), the Minister for Religious Affairs, of the orthodox Mizrahi Party.

The dominance of the two socialist parties, Mapai and Mapam, was fittingly reflected in the emergent state's main bureaucracies— in the Haganah and, afterwards, the IDF, in the Defence Ministry and its extensions (the military industries, the intelligence services, and so on), and in the other main ministries, the Foreign Ministry and the Treasury. Neither the religious nor the liberal parties enjoyed even a foothold in these main branches of government. Nor, save on certain economic issues, did they have much of a say in the policy- and decision-making, and policy implementation in the fields for which these bodies were responsible.

Among the minor parties, when it came to policy towards the Arabs, only the Revisionists spoke with a clear voice. The Revisionists continued through 1948 to lay claim to all Palestine west of the Jordan River, and to areas of Transjordan east of the Jordan, as the homeland of the Jewish people. They purported to believe that once the British left, Jew and Arab—under Jewish rule—could coexist in the homeland. In December 1947 they formally called on the Arabs to stay at peace. 'Only the British want war', argued the IZL.[1] The months of increasingly bitter warfare, however, persuaded the Revisionists that there could be no compromise over and coexistence in Palestine between Jew and Arab. As the breakaway Revisionist LHI group (*lohamei herut yisrael*, Freedom Fighters of Israel, called by the British the 'Stern Gang') put it in April, 'the Jewish Yishuv should not delude itself about good neighbourly relations and local peace agreements with the Arabs. All the

---

[1] JI, IZL Papers kaf—4/13/4, IZL poster 'To Our Arab Neighbours', 15 Dec. 1947.

agreements and treaties did not help the Hashomer Hatzair [Mapam] people in Mishmar Ha'emek . . . The front is throughout the country . . .'[2]

By April–early May 1948, with the life-and-death struggle at its height and an Arab invasion of Palestine imminent, a note of relief, satisfaction and even joy and gloating over the Arab exodus characterizes IZL broadcasts, though IZL commander Menachem Begin never openly espoused a policy of expulsion. But the IZL's goal with respect to the Arabs in Jewish territory was manifest in the attacks on Deir Yassin (9–10 April), Jaffa (late April) and in the Hills of Ephraim, southeast of Haifa (mid-May). LHI was franker. On 13 May, it declared:

A strong attack on the centres of the Arab population will intensify the movement of the refugees and all the roads in the direction of Transjordan and the neighbouring countries will be filled with panic-stricken masses and [this] will hamper the [enemy's] military movement, as happened during the collapse of France [in World War II] . . . A great opportunity has been given us, let us not waste it . . . The whole of this land will be ours . . .[3]

The other parties, of the religious and liberal persuasions, whether out of a sense of weakness or out of a feeling that what was happening was for the best, adopted a 'policy of non-intervention', of leaving Mapai and Mapam to decide what to do about the Arabs, and of letting the two socialist parties get on with the job. A sense of the thinking in the Progressive Party on the Arab question is afforded by the following debate in its executive committee in

[2] JI, LHI Papers, kaf—5/5/2, LHI broadcast on *Herut Yerushalyaim*, undated. LHI was composed of far-left / Communist and Revisionist factions, with the Communists continuing to believe in the possibility of fraternity with the Arabs and the Revisionists usually denying the possibility.

[3] JI, LHI Papers, kaf—5/5/2, LHI broadcast on *Herut Yerushalayim*, 13 May 1948. While Begin and the IZL leadership were careful not to openly espouse a policy of expulsion, it is clear that the IZL's military operations were designed with the aim of clearing out the Arab inhabitants of the areas they conquered. Following the massacre at Deir Yassin, the IZL fighters trucked out the remaining villagers to East Jerusalem. In May in the Hills of Ephraim the IZL assault ended in the flight of the majority of the villagers; and those who remained in place were, within days, swiftly sent packing in the direction of Umm al Fahm (then in the Iraqi-held zone in the West Bank). In their post-operational reports, especially after the conquest of Jaffa's Manshiya district, the IZL commanders emphasized their satisfaction with the fact that the assaults ahd precipitated mass civilian-Arab flight. The same note of satisfaction and accomplishment is rung in Begin's subsequent memoirs, *The Revolt*.

December 1948, just before the country's first general elections, regarding the party's attitude to the possibility of a refugee return:

> *Executive Committee member Avraham Riftin*: The Arab question; why should we now give an answer . . . Why should we be the first and only ones [to do so? A refugee return] is something very unpopular among the [Jewish] masses. Expansionism also now has great popularity. Why then should we now speak out against it . . . ?
>
> *Moshe Kol*. . . . [our attitude should be] not to now promise specific things, we do not know what our situation will be [in the future], perhaps we will have to bring here the Jews from the Arab countries and then certainly there will be no room here for the Arab refugees.
>
> *Moshe Tomkevich*. . . . We have no need to add minorities. Every modern state is moving towards eliminating minorities.
>
> *Yitzhak Golan*. We should not worry about bringing back the Arabs, who will be a Fifth Column in our state.[4]

But in the end only Mapai and Mapam really counted when it came to policy towards the Arabs. Together, they had the backing of more than half the electorate (as was demonstrated in the country's first general Knesset elections on 25 January 1949); almost all the kibbutzim were affiliated to one of the two parties; the powerful trade-union federation, the Histadrut—which both represented the workers, and owned and controlled major parts of the economy (banks, industrial plants, agricultural produce, and marketing)—was dominated by Mapai; the two parties dominated the cabinet; and, perhaps most important in the circumstances, they controlled and almost completely officered the defence establishment, which was responsible for the implementation of policy towards the Arabs. Indeed, the Haganah and, later, the IDF were peculiarly politicized in the sense that most officers and, indeed, units, 'belonged' to one of the two socialist parties. The sense of 'belonging' was especially pronounced in the units identified with Mapam (which included the three Palmah brigades and their officers).

In wartime, political parties tend to lose some of their importance; and party-political differences tend to blur. But in the peculiar circumstances of the Yishuv in 1948, Mapai's leaders and the Mapam party institutions carried especial weight. In Mapai, decisions on major defence and foreign-policy issues were discussed and taken

---

[4] Massuah-Kibbutz Tel Yitzhak Archive, 10/5/Mem, protocol of the meeting of the Progressive Party Executive Committee, 26 Dec. 1948.

by the party leader, Ben-Gurion, and his small circle of aides (Kaplan, Shertok, Eshkol (Shkolnik)), while party institutions abnegated responsibility in deference to the will of the leader. In Mapam, a party split down the middle by its constituent components (see below), the party institutions—Political Committee, Centre, Secretariat—were where policy was determined; the party had no leader.

MAPAI

Perhaps the most striking thing about Mapai and the Arab question in 1948 was that the party almost never discussed it. The party leaders apparently felt that the war and policy towards the Arabs with which it was inextricably bound up were the rightful domain of Ben-Gurion and of the Cabinet, the high-ranking army officers, and the upper reaches of the civil service. As these were controlled largely by Mapai personnel, the party institutions took the attitude that they should and could be allowed to operate with a minimum of party interference. It is likely that Ben-Gurion instructed the party stalwarts to avoid these subjects; debate inevitably would give rise to expressions of dissent. The bulk of the party officials apparently were satisfied with what was happening and agreed to the policy or the mode of thinking which to some degree was shaping developments.

Before 1948 Mapai's platform had called for the establishment of a Jewish state in any part of Palestine, whenever possible. The party continued to regard the whole land of Israel as by right the Jewish people's but insisted that Jew and Arab could and should coexist in it peacefully. Above all, Mapai policy, steered by Ben-Gurion, during the post-World War II years was opportunistic and flexible: the Zionist movement must be realistic and extract from history as much as history would allow at any given moment. The UN Partition Resolution of 29 November 1947 awarded the Jews a small state in Palestine: Ben-Gurion took what was offered.

But the outbreak of war radically changed things: the Jews had accepted a 'mini-state' in a divided Palestine; the Arabs had not, and had initiated hostilities to block the establishment of the Jewish state. The war afforded the Yishuv a historic opportunity to enlarge the Jewish state's borders and, as things turned out, to create a state without a very large Arab minority. The war would solve the

Yishuv's problem of lack of land, which was necessary to properly absorb and settle the expected influx of Jewish immigrants.

Ben-Gurion understood history, understood that war changed everything; a different set of 'rules' had come to apply. Land could and would be conquered and retained; there would be demographic changes. This approach emerged explicitly in Ben-Gurion's address at the meeting of the Mapai Council on 7 February: Western Jerusalem's Arab districts had been evacuated and a similar, permanent demographic change could be expected in much of the country as the war spread.[5] At that meeting, the party stalwarts, albeit briefly, for the first time touched on the Arab question, highlighting the two 'philosophies' which were recurrently to clash in the Yishuv during 1948. Eliahu Lulu (Hacarmeli), a Jerusalem branch leader, asked: who could believe that the Arabs of Palestine would long remain quiet and not rise up against the Yishuv? He, Lulu, had lived among them all his life. He knew them. The Arabs of Romema, a Jerusalem neighbourhood, were suing for peace. But this was only to 'save their houses . . . and no more'. Talk of possible Jewish–Arab fraternity was nonsense, he said.

By contrast, David Hacohen, from the Haifa branch, argued that though now the Yishuv was at war with the Arabs, eventually, 'we must not forget, we will also have to live with them. We talk with them peace and we must abide by our declarations . . . There is [Arab] terrorism and there is a Mufti . . . [But most of the Arabs] want to live quietly.'[6]

But for months thereafter, as the hostilities turned into full-scale war, Mapai's institutions refrained from discussing the war or the Arab question. They talked of party finances, enlarging the branches, relations with other parties, and preparations for setting up the civil bureaucracies of the new state. But not of the war and not of the Arabs. It was Golda Myerson (Meir) who broke the silence on 11 May, after a visit to the deserted quarters of Arab Haifa. Almost all the city's Arabs had fled over 22 April–1 May. Myerson spoke of the exodus, arguing that the party now had to face the issue, especially in terms of a possible return of some of the (friendly) villagers and preserving the abandoned villages for such a return. She also said the party had to determine Israeli behaviour towards

[5] David Ben-Gurion *Behilahem Yisrael* [as Israel fought] (Tel Aviv, 1952), 68–9.
[6] LPA 22/35, protocol of the meeting of the Mapai Council, Tel Aviv, 7 Feb. 1948.

the Arabs who had remained in the state: should it be such as to encourage or deter the exiles from returning? More generally, she noted that the Yishuv had made no plans concerning what to do with the population of conquered Arab towns and villages. 'I confess', she said:

and this may have been naïveté . . . but I believed that our army [would behave] not as all the world's armies [have behaved] . . . This is a slightly delicate and slightly complex problem. I advise the party Centre to discuss the problem soon, seriously and frankly . . . and according [to the conclusions reached] to instruct the [labour] movement as a whole . . . If the party doesn't take the initiative, then things won't be put right.

But, at the same time, Myerson cautioned against turning the issue —of the Yishuv's (and particularly the Haganah's) treatment of the Arabs—into a bone of inter-party (Mapai–Mapam) controversy. She enjoined the party to launch an education campaign in the ranks against looting.[7] But no other participants related to the Arab question, which was not on the agenda. Nor did Myerson's call for a full-scale party debate on the question in fact prompt such a debate. It was as if a large stone had been thrown into a pool—but had caused no ripples at all.

During the following two months the party institutions ignored the Arab question. Shertok (Sharett) made only a passing reference to it at the meeting of the Mapai Council on 18 June. The 'weakness' of the Palestine-Arab community came 'as a complete surprise . . . this community rooted in the country for generations, for hundreds of years . . . Who hoped, who could have anticipated that come war, it would pull up roots and wander . . . While we, who only recently arrived . . . and [taken] root . . . have displayed such steadfastness.' Shertok compared the exodus to the Sudeten-German exodus after World War II, and asked whether it wasn't just deserts for a war the Arabs had unleashed.[8]

The silence in Mapai was broken, briefly, by the impact of the expulsion from Lydda and Ramle on 12–13 July. The veteran, liberal Mapai Secretary General of the Histadrut, Yosef Sprinzak, at a meeting of the Histadrut's Executive Committee, said:

---

[7] LPA 48/23 aleph, protocol of the meeting of the Mapai Centre, 11 May 1948.
[8] LPA 36/22, protocols of the meetings of the 36th Mapai Council, Tel Aviv, 18 June 1948.

During [the past] weeks and months . . . positions [that is, attitudes or policies] have come into being which cause dissent among all of us in this house [the Histadrut] . . . This applies to the robbery, the looting and the behaviour [in general] of the Jewish occupation forces . . . It is possible that these things result from a logical imperative . . . It is possible there are security reasons. But . . . [there] must be a limit . . . [*Sprinzak cited the alleged expulsion from the village of Abu Ghosh and the 'evacuation' from Ramle*]. This situation is impossible . . . and the responsibility here is on all [the political leadership].

Eliezer Bauer (Be'eri), a Mapam figure from Kibbutz Hazore'a, at this point commented that the looting, which was not planned or centrally inspired but was not being curtailed, and the expulsions, taken together, testified that 'there is an unannounced but very effective intention that Arabs will not be left in the State of Israel'.

Eliahu Lulu (Hacarmeli) commented: 'I have taken part more than once in the argument about looting . . . It was once said that looting was one of the characteristics of the oriental [Jews]. [But] the Ashkenazim [western Jews] have superseded us also in this. Even in this we have been discriminated against . . .'. Another Histadrut Executive member (Mapai), Zvi Yehuda, from the agricultural settlement of Nahalal, noted that the Arab neighbourhoods of Haifa, among others, had been looted by kibbutz members from Sha'ar Ha'amakim, Ramat Yohanan, and Yagur: 'This is apparently a primitive instinct of *homo sapiens.*'[9]

The Lydda–Ramle episode and the increasing international pressure for a refugee return at last prompted a full-scale debate on the Arab question in the Mapai Centre. It took place on 24 July and was to be the only such debate in the course of 1948 in any of the party's institutions.

Sprinzak said: 'There is a feeling that *faits accomplis* are being created . . . The question is not whether the Arabs will return or not return. The question is whether the Arabs are [being or have been] expelled or not . . . This is important to our moral future . . . I want to know, who is creating the facts? And the facts are being created on orders.' There appears to be 'a line of action . . . of expropriation

[9] Histadrut Archive (Va'ad Hapoel Building, Tel Aviv), protocols of the meetings of the Executive Committee (*Ha'va'ad Hapoel*), 14 July 1948; and Yossi Amitai, *Ahvat Amim Bemivhan* (Brotherhood on Trial), (Cherikover, Tel Aviv, 1988), 38–9.

and of emptying the land of Arabs by force', said Sprinzak. If such was the policy, Sprinzak wanted to know, who had decided on it? He felt that the party institutions must first 'agree to it—I too have a right to a say in the matter', he said.

Sprinzak was supported only by Shmuel Yavnieli, one of the founders of Mapai and the Histadrut. Yavnieli described the question of a return as 'difficult'. In the past, he said, 'we resigned ourselves to the idea of a Jewish state with a large Arab minority'. Could Jews who 'for many generations were persecuted and expelled, slaughtered and destroyed', and were the butt of tyrants' contempt, now act like 'slaves who have become kings?' Mapai, he said, should declare that those Arabs who had not fought against the Yishuv could return. He cast doubt on the justice of previous transfers in history (Greeks and Turks), precedents cited by other Mapainiks in justification of the party line.

Ben-Gurion responded that one must pay attention to the chronology: 'The Arabs fled before the various places were conquered by the Jews.' Many of Haifa's, Jaffa's, and Jerusalem's Arabs fled before the Jews attacked. The same thing happened in Jenin. It was inexplicable, as if a 'dybbuk' had got into their souls. 'There was no necessity for them to flee', said Ben-Gurion. He added that when he had heard of the flight of Haifa's inhabitants, he had said: we will not run after them. 'If they flee—let them flee. We did not expel them.'

The two things that had surprised him during the war, said Ben-Gurion, were the Arab flight and the Jewish looting. 'It emerged that most of the Jews are thieves.' Everyone stole and looted, including 'the men of the [Jezreel] Valley, the cream of the pioneers, the parents of the Palmah [fighters]'.

Ben-Gurion, of course, had muddied the issue: he had implicitly denied that there had been expulsions by seemingly attributing the whole exodus to panic flight. Others at the meeting took a more forthright position. Shlomo Lavi, a veteran kibbutz movement and Mapai agricultural-sector leader, opposing a return, said: 'the . . . transfer of Arabs out of the country in my eyes is one of the most just, moral and correct things that can be done. I have thought this . . . for many years.' If a mass return were allowed, Lavi thought, the Arabs would 'stick a knife in [the nation's] back'. Lavi's views were endorsed by Avraham Katznelson. There is nothing 'more moral, from the viewpoint of universal human ethics, than the emptying of the Jewish State of the Arabs and their transfer elsewhere . . . This requires [the use of] force.' He, too cited the Sudeten precedent.

Lavi's statement had been in part a reaction to Shertok's political *tour d'horizon*. Shertok had said: 'It is desirable for us that the Arabs do not return, if it is at all possible . . . [This has] historical justification.' It is best for Israel and the Arab states in the long run that Israel should not have internal problems stemming from the existence of a large Arab minority, he implied. Shertok said, however, that he did not think the hour was ripe for this position 'to be formulated outwardly [that is, publicly]'.[10]

Sprinzak remained isolated. The matter was not raised again in the party's institutions for the duration of the war. Very few party members, if any, followed the example of Yitzhak Avira, a kibbutznik from Ashdot Ya'akov, in the Jordan Valley, who resigned his membership largely because of the party's stand on the Arab question. Not that he supported the return of the refugees, he wrote, but 'I do not agree with the party's way . . . I demand of our state that [we accord] elementary rights to the stranger living in our midst.'[11]

MAPAM

Mapam came into being in January 1948 when the Marxist Hashomer Hatza'ir and the socialist Hatnu'a Le'Ahdut Ha'avodah-Po'alei Zion parties united. Hashomer Hatza'ir, based largely on the kibbutzim of its Kibbutz Artzi Hashomer Hatza'ir kibbutz movement, had traditionally supported a bi-national Jewish–Arab state in Palestine, though gradually, after World War II, during the count-down to the 1947 UN partition resolution, it had come to accept the inevitability of partition, with the creation of a Jewish state in part of Palestine (albeit strongly linked by bonds of economy and 'anti-imperialism' to the other, Palestine-Arab state). Ahdut Ha'avodah, on the other hand, had traditionally supported the creation of a Jewish state in all of Palestine, with a Jewish majority coexisting with a well-treated Arab minority. Ahdut Ha'avodah, based largely

[10] LPA 48/23 aleph, protocol of the meeting of the Mapai Centre, 24 July 1948.
[11] LPA 1–48/3/4 aleph, Yitzhak Avira to the Mapai Centre, *c*.24 Oct. 1948.

on the kibbutzim of the Kibbut Me'uhad kibbutz movement, balked at Ben-Gurion's acceptance of the partition resolution and through the 1948 war pressed continuously for Haganah / IDF conquest of the whole of Mandate Palestine. The vision of what is today known as Greater Israel (*yisrael hashlema*, then known as *shlemut ha'aretz*) was the driving ideological force in Ahdut Ha'avodah and put it, perhaps, to the right of Mapai on foreign policy and defence issues (while it remained to Mapai's left in its economic and social outlook).

Against the backdrop of Jewish–Arab warfare in 1948, Mapam's leadership, equally drawn from the party's two component elements, was forced into an intricate and continuous juggling act to arrive at a succession of compromise formulas acceptable to their two (Hashomer Hatza'ir and Ahdut Ha'avodah) constituencies. Moreover, the realities of the conflict continuously and sorely tried the principles of the party members. As a Mapai kibbutz circular bitingly put it, not only was the 'distance' between the two Mapam components emerging starkly, but Hashomer Hatza'ir's behaviour was running contrary to its ideology. 'In practice, the kibbutzim of Hashomer Hatza'ir are participating in the acquisition of Arab property, harvest [Arab] fields, and more than this: [Hashomer Hatza'ir kibbutz] Mishmar Ha'emek was the first [kibbutz] to demand the destruction of the Arab villages around it . . .'. Meanwhile, stated the Mapai kibbutz circular, Mapam continued to blame Mapai for the Arab exodus.[12]

For Mapam's Hashomer Hatza'ir wing, the Jewish–Arab fighting was a large and continuing embarrassment. The party for years had hoped to turn the flank of Arab hostility and, making common cause with the 'peaceful, fraternal Arab workers', to achieve peaceful coexistence in a bi-national state. But national aspirations once again had overcome socialist premiss and vision. Hashomer Hatza'ir's left wing, led by the director of the party's Arab Department, Aharon Cohen, continued, however, to hope 'to build bridges to the Arab forces of progress . . .'.[13]

---

[12] Ihud Hakibbutzim ve'Hakvutzot Archive (Kibbutz Hulda) 111/2, *Igeret Gimel* (third circular), Gordonia-Maccabee Hatza'ir, the supreme leadership, Tel Aviv, 17 Sept. 1948.
[13] HHA-ACP 10.95.10 (6), Cohen to members of the Arab Department of Mapam, 17 Feb. 1948; and HHA 66.90 (1), protocols of the meeting of the Mapam Political Committee, 3 Mar. 1948.

But the war was destroying any chance of such fraternal bridge-building. In part, as Cohen saw things, it was a problem of Haganah policy—which appeared to him incapable of distinguishing between Arab foe and friend or potential friend: massive retaliation or 'aggressive defence' was killing any chance of peace. By March 1948, Cohen felt that there was a tragic lack of co-ordination between Haganah operations and the Yishuv's political purposes. He called for a full-scale debate in Mapam's Political Committee on policy towards the Arabs.[14] But no debate took place.

It took the shock of the exodus in late April of the Arabs of Haifa, a city where Jew and Arab had traditionally lived in peace, to stir Mapam's leaders sufficiently to agree to a full-scale debate.[15]

The internal Mapam debate on the Arab question began on 11 May and continued (in the Centre and in other party institutions) almost non-stop for the duration of the war. Marked by a deep and constant divide between the party's two wings, by less insistent disagreements within each wing, and by a comprehensive and often soul-searching treatment of the issues at stake, the protocols of the debate provide a major insight into the Zionist ethos and the Yishuv's views on the Palestine problem.

Preparatory to the start of the debate, Cohen wrote a memorandum entitled 'Our Arab Policy in the Midst of the War'. In his notes for the memorandum, penned on 6 May, Cohen wrote that 'a deliberate eviction [of the Arabs] is taking place . . . Others may rejoice—I , as a socialist, am ashamed and afraid . . . To win the war and lose the peace . . . the state [of Israel], when it arises, will live on its sword.'[16]

In the memorandum itself, Cohen lamented that during the first five months of the war, no basic discussion on policy towards the Arabs had taken place in the party. He cautioned that military victories should not go to the Yishuv's head. The Jewish state, once established, 'would have to fight for a long time for its existence. At best, our state will be an island in the large Arab sea.' Military successes had to pave the way for political accomplishments, or they

---

[14] HHA 66.90 (1), protocols of the meeting of the Political Committee of Mapam, 3 Mar. 1948; and HHA-ACP 10.95.11 (21), Cohen to Ya'acov Riftin and Leib Levite, 13 Mar. 1948.

[15] HHA 64.90 (1), protocol of the meeting of the Secretariat of the Mapam Centre, 27 Apr. 1948.

[16] HHA-ACP 10.95.10 (4), notes for memorandum, 6 May 1948.

were valueless. But, according to Cohen, 'out of certain political goals and not only out of military necessity' the Arabs were being driven out. 'In practice, a . . . "transfer" of the Arabs out of the area of the Jewish state was being carried out', and this would eventually redound against the Yishuv, both militarily (by increasing pan-Arab anger) and politically.[17]

Cohen's memorandum was distributed to the members of the Centre before the 11 May meeting. At the meeting, Mapam co-leader Aharon Zisling (Ahdut Ha'avodah) berated the Yishuv for the looting of Arab property but spoke rather vaguely on the fundamental issues of expulsion, a possible return, and the treatment of the Arabs. Co-leader Meir Ya'ari (Hashomer Hatza'ir) said he was shocked that the party's Haifa branch had done so little to deter the Arab exodus from the city; he too avoided the main issues. But several second-rank Mapam figures then turned to what was to become one of the foci of Mapam soul-searching—the role of the party's Haganah officers in the formulation and implementation of policy towards the Arabs. Yitzhak Yitzhaki called for a 'joint meeting' between the party's Arab affairs workers and its 'people [that is, officers] in the defence [establishment]', presumably to hammer out guidelines of behaviour towards the Arab population. Ze'ev Abramovich went further. The Secretariat of the Centre should inform Mapam's members serving in the defence establishment 'that the party will hold them responsible for preventing attacks on the Arab [civil] population . . .'.

Yitzhak Ben-Aharon, a leader of the Ahdut Ha'avodah wing, whose kibbutzim furnished a large proportion of the Palmah brigades' officers, was outraged.

They have loyally carried out their mission. No fact has been presented here [which speaks against] our comrades . . . Our comrades were 'responsible' in Mishmar Ha'emek, Haifa, Jerusalem, the South . . . [But] before the Centre has to comment upon [that is, condemn] our comrades, it must first discuss the facts, if there were any [cases of wrong-doing] . . . Anything else would be a motion of no confidence in our comrades.

Ben-Aharon differentiated between 'looting and robbery' (which he condemned) and expulsions out of military necessity.

[17] HHA-ACP 10.95.10 (4), 'Our Arab Policy in the Midst of the War', by Aharon Cohen, 10 May 1948.

How can one free [Kibbutz] Mishmar Ha'emek from being a military objective without uprooting [the neighbouring Arab] villages. There was no doubt about [the necessity] for this. Or: Along the road to Jerusalem. Safad. Haifa. And the result: The Arab fellah, even if he is tranquil, becomes a refugee . . . This may increase the number of refugees, etc. but it also increases our security . . .

Cohen, taking a conciliatory tack, suggested that instead of issuing a caution as proposed by Abramovich, the Political Committee should 'bring to the knowledge of the [Mapam members in the] defence forces [the sense of] the discussion [in the Centre]'. Later, the Political Committee and the officers should meet and discuss the matter.

Ya'acov Hazan, a (Hashomer Hatza'ir) Kibbutz Artzi leader, mollified Ben-Aharon by saying that the proposal had not aimed at 'casting aspersion on the officers . . .'. Hazan proposed that the Centre 'charge all our comrades in the Haganah with meticulously keeping to the party's positions on this question'.

The meeting in the end empowered the Political Committee to hold a clarificatory talk (*beirur*) with the Mapam officers, 'and to transmit instructions to our comrades in positions of command regarding a standard [pattern of] behaviour'.[18]

Given the military situation, which included the Arab invasion of Palestine on 15 May, Mapam proved unable to organize a meeting of its political leaders with 'its' Haganah commanders. But Ya'ari (Hashomer Hatza'ir) continued to insist that the party 'should instruct people who can affect the process of exodus or halt it'. The party should 'demand' that its members in the defence establishment discuss the matter with the Political Committee and, afterwards, proclaim its stand. Party co-Political Secretary, Leib Levite (Ahdut Ha'avodah), however, said that most of the commanders in any case couldn't come and it was best that the party's position be hammered out now, in the committee.

'We did not want . . . to expropriate [Arab property] . . . or anarchy . . . But there is a war and it has a logic [of its own] which must be carried through', said Levite. He spoke against the 'ambivalence' of at once calling for the restriction of the Jewish state to the partition borders and calling for expulsions and settlement by

<hr />

[18] HHA 68.9 (1), protocol of the meeting of the Mapam Centre, 11 May 1948.

Jews of the empty villages. Levite 'justified, supported, and endorsed
. . . every conquest and every eviction of every Arab settlement
necessitated by the war . . . Whoever wants peace, must win the
war . . .'. Looting, however, must be curbed with severity, he said.

The Hashomer Hatza'ir position at the 21 May meeting was
voiced by Aharon Cohen: 'What is being done today [is] that the
basis of existence of the Arabs who have left and will want to return
is being destroyed.' He implied that what was important was not so
much behaviour towards the conquered Arab communities as towards
the property and fields of those who had left: 'The crucial question is
the question of the return of the refugees', he said.[19]

The battle lines between the party's Hashomer Hatza'ir and
Ahdut Ha'avodah wings were being drawn. But they remained
blurred—mainly because all or almost all agreed that while the
Haganah had occasionally acted excessively and contrary to party
ideology, the Yishuv was fighting with its back to the wall; its
very existence was at stake. Now was not the time minutely and
rigorously to probe the ethics of every military action. But as the
immediate danger of Arab victory wore off, criticism of Haganah
treatment of Arab civilians mounted, inevitably highlighting the
Hashomer Hatza'ir–Ahdut Ha'avodah divide. The thrust of the
criticism was that Mapam's views appeared to have little or no effect
on the Yishuv's treatment of the overrun Arab communities, despite
the fact that Mapam's cadres made up much of the Haganah's
officer corps.

Eliezer Bauer (Be'eri) (Hashomer Hatza'ir) charged, at the
meeting on 26–7 May of Mapam's Political Committee, that under
the guise of military needs, things were being done for 'distinctly
political purposes, and without the party objecting and with party
members [carrying them out]'. Most of the Arabs who had left 'had
not been expelled. Many left before the [Haganah] conquests. The
conquests were a result of the flight.' But the Yishuv's attitude was
that almost every Arab who stayed 'had to be regarded as a spy or
potential saboteur'. At the same time, the economic infrastructure
of the Palestine Arabs was being destroyed, villages were being
razed, and new settlements on the abandoned sites were being
planned. The aim was 'to clear the State of Israel of Arabs . . .'.

[19] HHA 66.90 (1), protocols of the meeting of the Political Committee of Mapam,
21 May 1948.

No differentiation was being made between peaceful and hostile Arabs, and a policy of barring a return of the refugees was being implemented.

Moshe Erem said that the party had failed 'to educate the Haganah's units . . . Our comrades [if properly instructed] will serve as a brake on [expulsions] . . . We need commissars.' Erem thought that what was happening (Arab flight and the destruction of Arab villages) was happening without 'a directing hand'.

Hazan declared that a defeat in the war would mean 'the end of the Jewish people'. Hence, the war had to be fought with all necessary 'brutality and efficiency'. But, the fighting aside, what was happening was 'a political, moral, public and socialist scandal'. He specifically denounced the way the Haganah was treating the Arabs who had stayed put and the destruction of the villages. He spoke of Abu Shusha, a village near his home kibbutz, Mishmar Ha'emek. The Haganah had completely bulldozed the village instead of distinguishing between houses of enemies and houses of friends of the Yishuv. He spoke of Haganah 'killing, robbery, rape. I don't think our army should be like every army'. Hazan concluded by rejecting the establishment of a Jewish settlement on the lands of Abu Shusha. '[I] accept that next to us [that is, next to his kibbutz, Mishmar Ha'emek] there will always be an Arab village', he said.

Ya'acov Amit (Hashomer Hatza'ir), of Kibbutz Beit Zera, south of the Sea of Galilee, said that it was all very well to blame Ben-Gurion for everything. But why wasn't Mapam's dominant presence in the army being translated into political praxis? He asked, specifically, why Israel Galili, of the Haganah General Staff, was keeping quiet and why Hazan, on the board of directors of the JNF, was not curbing Yosef Weitz, the JNF Lands Department director, who was organizing the destruction of Arab villages. 'What are you sitting [on these bodies] for? Either intervene or don't sit there.'

Eliezer Pra'i (later, Peri) (Hashomer Hatza'ir), the editor of Mapam's daily, *Al Hamishmar*, questioned whether there was always a military reason for the razing of the Arab villages. He could understand, he said, the 'joy of the comrades who have been freed from the fear of the [neighbouring] Arab villages'. But what was happening was undermining 'the moral and political base on which our party was founded . . . Among the best of our comrades the thought has crept in that perhaps it is possible politically to achieve our ingathering in the Land of Israel by Hitlerite-Nazi means.'

But a number of committee members—Levi Kantor (Ahdut Ha'avodah) and Haim Darin-Drabkin—objected to what amounted to a wholesale condemnation of the Yishuv's policy towards the Arab population. Kantor said that if Mapam was to act as Pra'i was recommending, 'it would be seen as sabotaging our war effort'. Darin-Drabkin, a Tel Aviv party member, dismissed this facile condemnation of expulsions by pointing out that had Jaffa's population remained *in situ*, it would have 'posed a difficult problem' for the Haganah. The same applied to the Arab villages around Mishmar Ha'emek. 'Because of the desire to destroy [the Yishuv] on the part of the Arabs, there [cannot be] here Geneva conventions regarding civilian population.' Darin-Drabkin denied that there was any centrally organized policy of expulsion.[20]

Thinking critically and publicizing criticism of aspects of military operations while the war was in full swing was difficult. The commencement of the First Truce (11 June) opened the floodgates of dissent. Mapam co-leader Ya'ari let loose on 14 June, in an address to Kibbutz Artzi military men: 'There are those who say . . . "we did not expel them. They fled of their own accord . . . and so why should we not inherit after they expropriated themselves, as it were. Why should we not clear the area [of Arabs] and not snatch this unheard of opportunity?"' But these are 'barbarous growths', said Ya'ari; 'anti-imperialism and racism' cannot dwell together. 'In truth, thousands did flee, but not always of their own will. There were shameful episodes. . . . There was no necessity for all the villages to be emptied . . .'.

Ya'ari, however, opposed the return of the refugees during the hostilities.[21] So did Agriculture Minister Zisling (Ahdut Ha'avoda). But Zisling warned, at the Cabinet meeting of 16 June, that the Yishuv was on a course that would endanger any possibility of making peace. 'Hundreds of thousands of Arab dispossessed . . . are growing into haters who will promote war against us throughout

[20] HHA 66.90 (1), protocol of the meeting of the Political Committee of Mapam, 26–7 May 1948.
[21] HHA *Yediot Hakibbutz Ha'artzi-Hashomer Hatza'ir*, no. 274 (43), 8 Aug. 1948, 'If You Go to War', a speech by Meir Ya'ari to Kibbutz Artzi conscripts, 14 June 1948.

the Middle East . . . They will bear in their breasts the desire for revenge, compensation, and a return . . .'. The Arabs should be allowed to return to certain sites, such as Beisan, whose houses should not be destroyed. Zisling, joining Minority Affairs Minister Shitrit, sharply attacked Ben-Gurion for the destruction of the Arab villages. But when it cut close to the bone, Zisling spoke otherwise. 'I do not want the Arabs to return to Qumiya', he said, speaking of the abandoned village overlooking his own kibbutz, Ein Harod, in the Jezreel Valley.[22]

Zisling's criticism of Ben-Gurion was prompted by the sense of the meeting of Mapam's Political Committee of 15 June. Cohen had charged that 'it had depended on us whether the Arabs stayed or fled . . . [They had fled] and this was [the implementation of] Ben-Gurion's line in which our comrades are [also] active.' Ya'ari had pointed to the 'moral contradiction between [our ideology and the fact that] our comrades are implementing [Ben-Gurion's policy] . . . The comrades are doing things contrary to their conscience because they received orders from above. They also do things on their own initiative.' Ya'ari proposed that the party inform its officers in the IDF 'that they should carry out every order [even if it] runs against their conscience but should inform their superiors that they were acting against their conscience'. Ya'ari also proposed that the party probe the actions of several commanders. He was supported by Mordechai Bentov (Hashomer Hatza'ir), the Minister of Labour.

The Political Committee, in an eleven-point policy statement that summed up the prolonged party debate on the Arab question, ruled that 'the party opposes the objective of expelling the Arabs from the areas of the emergent Jewish state'; proposes that the government 'call on peace-minded Arabs to stay put . . .'; opposes 'the destruction of Arab settlements . . . not out of military necessity'; and proposes that the government call on 'the Arab citizens of the Jewish state, come peace, to return to a life of peace, respect and creativity', and that property be restored to the returnees.[23]

[22] KMA-AZP 9/9/3, transcript of Zisling's statements at the Cabinet meeting of 16 June 1948.

[23] HHA 66.90 (1), protocols of the meeting of the Mapam Political Committee, 15 June 1948; and HHA-ACP 10.95.11 (1), 'Our Policy towards the Arabs during the War, (decisions of the Political Committee from 15 June 1948)', the Secretariat of the Mapam Centre, Tel Aviv, 23 June 1948; a circular to party activists.

The Political Committee debate on the exact wording of the policy statement highlighted the growing Hashomer Hatza'ir–Ahdut Ha'avodah divide. The debate over the clause regarding the destruction of Arab villages was especially heated. Pra'i and Bauer sought a blanket prohibition of such destruction. Ben-Aharon demanded—and got—the qualification allowing for destruction 'out of military necessity'. The wording of the clause on a refugee return was also strongly debated. Zisling opposed the phrase in the original draft concerning the return of the refugees 'to their [original] places'—Qumiya, for example. The phrase was duly excised from the policy statement.[24]

Throughout June and July, Mapam spokesmen persistently criticized and condemned the continuing destruction of Arab villages.

The Israeli offensives of the 'Ten Days' (9–18 July) and the expulsion of the inhabitants of Lydda and Ramle substantially sharpened, indeed qualitatively transformed, the rumblings within Mapam to open, across-the-board dissent from Mapai over Arab policy. For the first time, the possibility surfaced of Mapam, even in mid-war, breaking with Mapai and bolting the coalition. On 15 July, Hazan said: 'The robbery, killing, expulsion, and rape of the Arabs could reach such proportions that we would [no longer] be able to stand it. But even then we would have to think [twice about bolting the coalition].' Hazan proposed, rather, that the party try to influence the course of events from within.[25] Zisling's criticism of Ben-Gurion and of the Lydda–Ramle expulsion in the Cabinet meeting of 21 July was bitter; Hazan, at the 14 July meeting of the Histadrut's Executive Committee, was only marginally less scathing about what he defined as the Mapai line.[26]

At the same time, the Lydda–Ramle expulsions dramatically conjured up and highlighted the divide within Mapam between its two component elements, inevitably renewing the full-scale debate on the party's Arab policy. The debate variously focused on ethical, political, and strategic considerations. The expulsions from the two towns were generally condemned by the Hashomer Hatza'ir stalwarts,

[24] HHA 66.90 (1), protocol of the meeting of the Mapam Political Committee, 15 June 1948.
[25] HHA 68.9 (1), protocol of the meeting of the Mapam Centre, 15 July 1948.
[26] KMA-AZP 9.9.3, transcripts of Zisling's statement in the Cabinet meeting of 21 July 1948; and Histadrut Archive (Va'ad Hapoel Building), protocol of the meeting of the Executive Committee, 14 July 1948.

as was the behaviour and judgement of the commanders involved, principally OC Operation Dani Yigal Allon (Ahdut Ha'avodah). The Ahdut Ha'avodah wing tended to support the generals, though eventually they acquiesced in a compromise formula which criticized the generals for expelling the inhabitants of Ramle but condoned the expulsion from Lydda (because, it was argued, the Lydda inhabitants had 'rebelled' against the IDF after conquest whereas Ramle's population had not). The compromise was reached after an informal 'investigation' of the generals by the party leaders, which included a 'guided tour', with explanations, by Allon of a handful of the leaders around the Operation Dani battlefields even before the operation had ended. While arguing strategic necessity, Allon pointed to Ben-Gurion as the origin of the expulsion order. The debate within Mapam about Lydda and Ramle sputtered on for months, the twin expulsion emerging as a general symbol of the alleged, continuing 'transfer policy' (orchestrated by Mapai) and, more particularly, of Mapam's involvement, through 'its' generals, in that policy.[27]

Ben-Aharon (Ahdut Ha'avodah) denied that there was a Hashomer Hatza'ir–Ahdut Ha'avodah divide over Arab policy. 'There is no argument over our attitude to the Arabs. There is no argument over our opposition in principle, fundamentally, to the expulsion of Arabs from the country . . . [and] against the destruction of [their] property . . .'. The argument, he said, was about 'what constitutes military necessity'. The problem lay in differentiating between expulsions out of military necessity and out of 'military convenience. We are against the expulsion of Arabs and the destruction of villages out of convenience . . .'. But there was no argument in the party over expulsions prompted by military necessity, said Ben-Aharon.[28]

Hashomer Hatza'ir agreed that there were cases of military necessity. The executive committee of the Kibbutz Artzi in August, in a bulletin to its members in the IDF, acknowledged that 'there are cases where the army is interested in evicting the Arabs after

---

[27] For further detail on the political debate within Mapam and between Mapam and Mapai on Lydda and Ramle, see Benny Morris 'Operation Dani and the Expulsion from Lydda and Ramle, July 1948', in *Middle East Journal*, Winter, 1986.

[28] HHA 10.18, transcripts of speeches delivered at the meeting of Mapam defence activists, 26 July 1948.

conquering an Arab site because we do not have enough forces to leave a garrison behind . . . The army cannot advance when to its rear villages or towns remain without effective supervision.'[29]

But, by and large, the Hashomer Hatza'ir leaders did not see the question as one of military necessity versus convenience: what was happening was 'political', in terms both of motivation and consequence. The expulsions and the razing of the villages were being carried out for political reasons and, often, by Ahdut Ha'avodah-affiliated officers (Yigal Allon, OC Palmah and OC Operation Dani; Yitzhak Rabin, Palmah OC Operations and Operation Dani OC Operations; and Moshe Carmel, OC Northern Front). As Hazan put it, concerning the destruction of the villages one had to distinguish between what occurred during battle and what occurred four months later. And 'there were many cases of expulsion that were not a necessity of war . . . If we adopt [Ahdut Ha'avodah Mapam Centre member Menahem] Durman's approach [that is, the fewer Arabs left in Israel, the better], it is possible to conclude that it is necessary to expel all the Arabs from the country', said Hazan. Expelling the Arabs will create enemies 'for at least two generations. There is no future for Zionism if one assumes frequent wars.'[30]

Ahdut Ha'avodah's stance was clarified by Yosef Tabenkin, the commander of the Palmah's Harel Brigade and the son of Yitzhak Tabenkin, Ahdut Ha'avodah's leader, at the meeting of Mapam defence figures on 26 July. The young Tabenkin did not agree to a transfer policy, he said. After the war the Arabs would return to the country and 'the land would not be clear of Arabs. In this country Arabs and Jews will live on foundations of equality.' But, and here was the crux, 'in war—do as in war'. Tabenkin defended the expulsion from Ramle by saying that he 'did not want the "just men" [that is, population] of Ramle [left] behind us, on the road to Jerusalem, now. The just men of Ramle . . . would have constituted a danger had they still been in Ramle . . .'.[31]

The Mapam (and Shitrit) attacks on Ben-Gurion, the Mapai 'line'

[29] HHA 11.18, 'Bulletin to the Mobilized [Members] no. 3', the executive committee of the Kibbutz Artzi, 25 Aug. 1948.

[30] HHA 68.9 (1), protocol of the meeting of the Mapam Centre, 15–16 July 1948.

[31] HHA 10.18, transcripts of the speeches at the meeting of the Mapam defence activists, 26 July 1948.

and the destructive activities of the Haganah / IDF and Weitz in
May–July 1948 had had a major effect. It had resulted in the curtailment and cessation of the First Transfer Committee's operations
and in the issuing of the order of 6 July by IDF Deputy Chief of Staff
General Zvi Ayalon (in the name of the Chief of Staff) forbidding
destruction of villages and expulsions without specific permission
from Ben-Gurion. And it had induced Ben-Gurion and the generals
to have to 'think twice' in the future when contemplating such acts
and, if decided upon, to carry them out without fanfare or, usually,
explicit, written orders. Aharon Cohen in July felt able to write that
he discerned 'a certain measure of reward to [our] stubborn and
occasionally desperate efforts'. He was speaking of the non-expulsion
of the Arab inhabitants of Nazareth and its environs in mid-July, of
the ministerial interventions that stopped the expulsion of the
inhabitants of Al Fureidis and 'Arab al Ghawarina (Khirbet Jisr az
Zarka) in the Coastal Plain and to the appointment of a string of
Mapam officials to key Arab-affairs posts in the bureaucracies
(Moshe Erem in the Minority Affairs Ministry, Eliezer Bauer in the
Labour Ministry, Yitzhak Yitzhaki in the Histadrut's Arab Department, Yosef Vashitz in Haifa's Arab-Affairs Committee, and so
on).[32]

But during the late summer and autumn of 1948, the Hashomer
Hatza'ir–Ahdut Ha'avodah divide became somewhat blurred and
irrelevant as many of the party's kibbutzim, of both wings, cast
covetous eyes on the abandoned Arab lands. New settlements,
affiliated to both the Kibbutz Me'uhad and Kibbutz Artzi movements,
were set up on these lands.

A rearguard action was fought by some Hashomer Hatza'ir activists
to halt the process. The protests focused not on stopping the
establishment of new settlements or on stopping the cultivation of
Arab fields (both of which would have been regarded as unpatriotic)
but on halting the destruction of the empty villages (in the interest
of a possible return and future Arab–Jewish fraternity). Letters
denouncing the destruction reached the party leaders (and IDF and
civil bodies) from a number of kibbutzim. These acted as 'constituency' grass-roots pressure on the leaders in internal party
deliberations and in the set-tos with Mapai.[33]

---

[32] HHA-ACP 10.95.10 (4), Cohen (Kibbutz Sha'ar Ha'amakim) to Ya'acov
Mayus (Prague), 30 July 1948.
[33] See e.g. HHA-ACP 10.95.10 (6), Peterzil to Erem, Bentov, Hazan, Zisling, 10

There were also some direct attacks on the take-over of Arab lands and the establishment on them of new settlements. Baruch Lin, of Kibbutz Mishmar Ha'emek (Hashomer Hatza'ir), a Kibbutz Artzi leader, denounced what most of his party comrades were beginning to call an 'agrarian reform' as nothing other than a 'transfer'. 'Transfer, not agrarian reform, is now in the air', he said. It was being argued that new settlements were being established on sites out of 'strategic need'. 'But every Arab village in the country [could be settled on the basis of] "strategic need"', he said. Most Mapam members, including men like Hazan and Erem, that autumn supported what they called 'agrarian reform', which meant the establishment of new Jewish settlements on half the abandoned Arab lands, with the other half (or less) to be set aside for the possible return of the refugees. It was argued that 'intensification' of cultivation would mean that half the lands could produce more than all of them had before 'intensification'. Even Hashomer Hatza'ir stalwarts such as Aharon Cohen, given the kibbutzim's avid desire for more land, refrained from coming out against the proposed agrarian reform. At meetings of the Political Committee, Ahdut Ha'avodah leaders Yitzhak Tabenkin and Ben-Aharon strongly supported the 'reform' but resisted setting aside lands for returnees near new Jewish settlements. 'Otherwise', argued Ben-Aharon, 'we will establish a settlement with a defect, which will constantly struggle with the [neighbouring] Arab settlements over every dunam of land.' Ben-Aharon demanded the establishment of new settlements—'as a strategic-settlement necessity'—in the Jerusalem corridor. The corridor's original inhabitants should be 'settled elsewhere'. Tabenkin, for his part, said: 'It is necessary to settle . . . if we don't settle on the [abandoned] land, we will endanger our existence . . . we must not exhibit a guilty conscience in this matter . . . We must cultivate every [piece of land] without determining in advance what will be the fate of this settlement [drive] when the Arabs return.'[34]

The Mapam kibbutz members' desire for new lands was such that it touched not only those tracts abandoned by fleeing Arabs but

Aug. 1948; and HHA-ACP 10.95.10 (5), Kibbutz Sha'ar Ha'amakim to Golani Brigade HQ, 8 Aug. 1948.

[34] HHA 66.90 (1), protocol of the meeting of the Mapam Political Committee, 19 Aug. 1948.

also fields belonging to Arabs still in the country. Yagur member Noah Prover, who served as an inspector of abandoned property in the Haifa area, complained that kibbutz members 'are expelling ... Arabs who have legal rights or rental contracts regarding lands near old and new kibbutzim . . .'.[35]

During August–October 1948 the debate between Mapam's constituent factions, and within Hashomer Hatza'ir itself, regarding the need to set aside lands for possible returnees increased in volume.[36] A yawning gap developed between Mapam's political professions and its members' actions. The party—and more vehemently, the Kibbutz Artzi—supported a refugee return. But the kibbutzim coveted the abandoned lands. The kibbutzim took over fields, and some kibbutzim egged on expulsions and helped to destroy villages.[37]

In September, Yitzhak Tabenkin, the Ahdut Ha'avodah and Kibbutz Me'uhad leader, presented what amounted to his credo on the Arab question, the refugees, and the lands issue. It is worth quoting at length:

We are on the eve of [the renewal of] the hostilities ... Among us there was an argument whether to expel or not. [But] the Arabs did not even ask our people whom they caught and killed barbarically . . . Every expulsion is difficult. But if we are confronted by the choice [:] the expulsion [of Arabs] or the murder [of Jews] everyone will choose expulsion . . . Though [the expulsion from] Ramle does not weigh on my conscience, one might say that most of the Arabs [in the town] were not expelled but were tricked [into leaving]. There are some 300,000 Arab refugees and had they known that the result of the war would have been their expulsion it is possible that they would not have chosen [this course] . . . Are our [Jewish] refugees worse than Arab refugees? . . . I am not indifferent towards the Arab refugees, but we are a people of refugees . . . And *aliya* [that is, Jewish immigration to Israel] is inconceivable without land . . . I accept the Arab's

---

[35] Amitai, *Ahvat Amim Bemivhan*, p. 43.

[36] See e.g. HHA 5.10.5 (2), protocol of the meeting of the executive committee of the Kibbutz Artzi, 9 Sept. 1948; HHA 66.90 (1), protocol of the meeting of the Political Committee of Mapam, 25 Nov. 1948; and KMA 12–3, protocols of the meeting of the Kibbutz Me'uhad Council, statement by Reuven Cohen, 6 Nov. 1948.

[37] See e.g. CZA A246–13, p. 2,354, entry for 14 Apr. 1948, p. 2,364, entry for 23 Apr. 1948, p. 2,367, entry for 26 Apr. 1948, for the case of Kibbutz Kfar Masaryk and the Ghawarina beduins of Haifa Bay; HHA-ACP 10.95.10 (6), Salem Horowitz to Aharon Cohen or Mapam's Political Committee, Sept. 1948, requesting the destruction of a string of Lower Galilee villages; and HHA-ACP 10.95.10 (5), Peterzil to Erem, Bentov, Hazan, and Zisling, 10 Aug. 1948, on destruction by Kibbutz Ma'ayan Baruch of a neighbouring Arab village.

right to live here, a right equal to our own. But we need not announce this
now . . . He who fights against us now endangers his rights . . . Our war is
defensive . . . I do not regard the Ramle affair as embarrassing to us . . .
The Arab will be happiest [living] among [fellow] Arabs . . . Ramle was not
a failure for our party . . . The failure of our party is that we did not know
how to resist one man's [Ben-Gurion's] assumption of control over the
army and the war effort . . .[38]

The renewed fighting in October in the Galilee and the Negev was
swiftly followed by reports and rumours of IDF atrocities. Mapam
commanders and soldiers were implicated, directly and indirectly.
Months before, Ya'ari had lamented what he saw as the failure of
the Kibbutz Artzi's education system: it had produced, when it
came to fighting, a 'wonderful generation, brave and courageous'.
But, on the moral plane, the citizen-soldiers had not stood the test
of the war.

The youth we nurtured in the Palmah, including kibbutz members, have
[occasionally] turned Arabs into slaves; they shoot defenceless Arab men
and women, not in battle . . . Is it permissible to kill prisoners of war? I
hoped that there would be some who would rebel and disobey [orders] to
kill and would stand trial—and not one appeared . . . They are not against
transfer. What does it mean . . . to empty all the villages? . . . What did we
labour for . . .?

Ya'ari asked the kibbutz educators.[39]

The atrocities of late October in Operation Hiram (in the Galilee)
and Operation Yoav (in the south) confronted Mapam with the
practical problem of persuading 'its' officers and soldiers to give
evidence before the ministerial commission of inquiry (in the end
carried out by Attorney-General Ya'acov Shimshon Shapira alone)
that had come into being partly because of Mapam's pressure on
Ben-Gurion. The atrocities, and Mapam's part in them, were
debated in the party's Political Committee on 11 November. The
Hashomer Hatza'ir–Ahdut Ha'avodah divide again surfaced. Aharon

---

[38] HHA 68.9 (1), protocol of the meeting of the Mapam Centre, 16 Sept. 1948.
[39] HHA-Meir Ya'ari Papers 7.95.3—aleph (2), text of speech by Ya'ari at a
meeting of teachers in Kibbutz Mizra, August 1948. Ya'ari was misinformed. There
had been a number of cases of soldiers refusing to carry out barbaric orders. At least
two cases—of soldiers of the Palmah's Third Battalion outside 'Ein az Zeitun and
soldiers of the 89th Battalion, 8th Brigade at Ad Dawayima—are recorded in Benny
Morris, *The Birth of the Palestinian Refugee Problem, 1947–1949* (Cambridge,
1988).

Cohen, speaking of the divorce in Mapam between theory and praxis, called for a full-scale, internal party investigation. Galili (Ahdut Ha'avodah) said that the Mapam leaders should not rush to accuse party members without proof. Bentov (Hashomer Hatza'ir), the Labour Minister, said that the ministerial probe was encountering difficulties as the troops were refusing to testify. Benny Maharshak (Ahdut Ha'avodah), a senior Palmah officer, said that Mapam members should not use phrases like 'Nazi acts' to describe IDF actions. Ya'acov Riftin, a Hashomer Hatza'ir hardliner, said that there was no need to 'mix up' the subject of expulsions with atrocities; he implied that there was no connection between the two. But he called for 'death sentences' for the perpetrators of atrocities. The committee decided that it should hold an investigatory meeting (*beirur*) with the relevant Mapam commanders.[40] At the same time, the executive committee of the Kibbutz Artzi instructed the movement's soldiers to come forward and testify on IDF maltreatment of Arabs in the recent campaigns.[41]

During October–November, various Kibbutz Artzi–Hashomer Hatza'ir members began publicly to criticize Abba Kovner, of Kibbutz Ein Ha'Horesh, a well-known poet and World War II partisan and Holocaust survivor. Kovner served as the cultural officer in the Givati Brigade and composed the brigade HQ's circulars to the troops. During the July fighting, addressing himself to the brigade's armoured-vehicle drivers, Kovner had written: 'Don't be deterred, sons: dogs of murder—their sentence must be blood! And the better you run over these dogs of blood the better will you love the beautiful, the good, and freedom . . .'.

On 20 October, Kovner wrote: 'The bayonet is just and the blood is free, because the vision of retribution says: Vengeance. Vengeance. Vengeance!'

Kovner was personally assailed in various kibbutz gatherings and in Mapam party meetings. In a letter to the Mapam leadership, Kovner denounced his critics and even charged that one of them, a senior party figure, had called him 'a Fascist propagandist'. Kovner demanded that the charges against him be publicly aired in a

[40] HHA 66.90 (1), protocol of the meeting of the Mapam Political Committee, 11 Nov. 1948.
[41] HHA 5.18 (1), 'Bulletin no. 76', the executive committee of the Kibbutz Artzi, 12 Nov. 1948.

meeting of the party institutions.[42] It seems that such a meeting did not take place.

The Kibbutz Artzi's annual council, held in Nahariya over 10–12 December and gathering together representatives from all the movement's kibbutzim and the Hashomer Hatza'ir leadership, turned into a full-fledged and painful debate on Mapam's policies during the war. Everything was reviewed—the Haganah's initial retaliatory strategy, the offensives of April, Haifa, expulsions, the destruction of the villages, Lydda–Ramle, relations within Mapam, relations with Mapai, the takeover of lands, the establishment of new settlements. It was a prolonged stock-taking and a summation.

Many of the kibbutzim held preparatory general meetings of members before the council convened. Often, the focus was the refugee problem. 'In every kibbutz the Arab question gives no rest', one leading Kibbutz Artzi member reported.[43]

At Ma'anit (Narbata), east of Hadera, the members met in early December.

There are contradictions between the declared positions of the Kibbutz Artzi and the feeling of some of the [kibbutz] comrades . . . We have comrades who are close to Mapai's positions on the Arab question. This is a serious danger as one cannot long be a revolutionary socialist outside and a reactionary at home. It is difficult to withstand the waves of chauvinism which arise during times of victory . . . And those comrades who think that it is possible to develop a democratic, free state while badly treating the Arabs are mistaken,

said Yosef Menahem, one of the kibbutz leaders.

Binyamin Winter, the *mukhtar* (headman) of Ma'anit, said that during thirty years of struggle, 'we failed' to persuade the Arabs that 'they could live with us in peace'. The British, on the other hand, succeeded in convincing the Arabs of the 'slaughter' the Jews would unleash against the Arabs if the Jews took control. So the Arabs fled, fearing 'retribution'. Now, the 'transfer' was a *fait accompli*, and there would be no return. Winter asserted that 'our efforts' had been responsible for the fact that the inhabitants of Al

[42] Avraham Eilon, *Hativat Givati Mul Hapolesh Hamitzri* (The Givati Brigade Opposite the Egyptian Invader) (IDF Press-Ma'arachot, 1963), 298–9, 524; and Amitai, *Ahvat Amim Bemivhan*, 43–4.

[43] HHA 5.20.5 (4), protocols of the Kibbutz Artzi council, statement by Zvi Lubliner, 12 Dec. 1948.

Fureidis and 'Arab al Ghawarina had not fled. 'Elswhere we did not manage to hold the Arabs. Now, the Right also regrets the expulsion of the Arabs—as it lacks unorganized labourers. But the matter was primarily strategic. Had we left the villages intact and made do only with evicting the young men, we would have endangered our security.' Winter dismissed the notion that Mapai was responsible: true, 'Ben-Gurion got his way'. But 'the commanders were ours [that is, Mapam's]'.

Another kibbutz member, 'Lucy', said that even had Mapam controlled the Yishuv, 'things would not have been different. We would have demolished like the others. One can do no other [in wartime] . . . For us every Arab is now a hater or potential hater . . . He who wishes to kill you, rise up first and kill him.'

Pinhas Ger took a contrary tack. Zionism had never intended 'to settle the Jewish immigrant in the expelled Arab's house'. He upheld the Arabs' right to return and thought that Jewish housing problems should not be solved at the expense of the Arabs. But other kibbutz members opposed a return: 'In Transjordan and Iran [Iraq?] there are wastelands [where the refugees could be resettled]', said Yardena Avner.[44] This was to be a microcosm of the following, wider Kibbutz Artzi council debate.

In advance of that debate, Ya'ari wrote that the Yishuv had failed to exploit the opportunity offered by the IDF victories. 'We could have stopped the confused flight and ended the panic that took hold of the whole Arab population . . . We could have nurtured among them allies . . . Instead . . . we reinforced the panic. Through our behaviour we turned the Arab flight into an almost complete exodus.'[45]

The debate of 10–12 December was frank and comprehensive. Eliezer Hacohen, of Kibbutz Beit Alfa, summarized the year that had passed by saying that, on the Arab question, Mapam had in effect been 'paralysed' and its platform ignored. The fault lay with the 'party's' Haganah / IDF officers, who had carried out operations which ran counter to the party line. He seemed to imply that the party had not disciplined the commanders because of the Hashomer

[44] Ma'anit Archive 4/ayin, *Benirbata* (the kibbutz's internal bulletin), 8 Dec. 1948.

[45] HHA 5.20.5 (3), Meir Ya'ari, 'The Kibbutz Artzi after the [Mapam] Unification and after the Establishment of the State', Nov.–Dec. 1948.

Hatza'ir–Ahdut Ha'avodah divide, which could have turned into a schism had the issue been strongly pressed by the Left.

Ya'ari went so far as to accuse his Hashomer Hatza'ir colleague Riftin of taking the side of the Ahdut Ha'avodah generals, specifically over the expulsion from Ramle, 'in order to maintain party unity'. So, while the party officially repudiated a policy of expulsion, 'its' generals had helped implement it, and Ahdut Ha'avodah's leaders and some Hashomer Hatza'ir figures had sanctioned it.

Ya'ari said that in the early months of the war, when the Yishuv fought with 'Stens [that is, sub-machine-guns] against cannon . . . it was possible to understand' what had happened. But later, from Lydda–Ramle onwards, and despite party decisions to the contrary, the generals had done things with which Mapam could not agree. But the party leaders had also been remiss—in having failed to punish the miscreants.

As to the present and future, interest focused on the issue of a refugee return. In principle, all the Council participants favoured a return. But the party was 'afraid to speak up', said Zvi Lubliner. General elections (eventually held on 25 January 1949) were at hand and the party's opponents, it was feared, would argue that Mapam 'wanted to bring back Arabs' at the expense of new Jewish immigrants. Allowing refugees back was unpopular; advocating it would cost Mapam votes. None the less, Hazan spoke up in favour of a return. If there was no return, Israel would be encircled by 'a whole generation [seething with] hatred . . . ready at any moment for a war of revenge, a generation that will view us as robbers and murderers, who stole from it everything—its homeland and its property'. Riftin, too, backed a return, but not necessarily to the original abandoned sites, and 'within the framework of a development plan'. But party member Hemda Gilrovitz suspected that conditioning a return on a 'development plan' would result in no return. Weitz, she said, also spoke originally of a return within the framework of such a plan. She added that the party didn't seem to be 'exerting itself' too much on behalf of a return. David Kna'ani, of Kibbutz Merhavia, supporting a return, said that 'the Land of Israel is [the Arabs'] home as it is ours'.

Outright opponents of a return were rather mute at the council meeting. A member of Kibbutz Sha'ar Hagolan, identified in the protocols only as 'Shmuel', was probably the most forthright:

For us, the question of the refugees is not abstract. We see the long line of huts of the refugees [across the Jordan River] from our observation posts. There are many dangers in allowing back the refugees. I am not happy with the idea of having a large Arab minority. Our treatment of the Arabs is not so bad. We were told here of an example of transfer from Greece [to Turkey]. This transfer brought Greece thirty years of peace . . .

The Kibbutz Artzi council concluded its meeting on 12 December by adopting a number of resolutions, which included allowing the return of 'the peace-minded refugees', and their absorption within a 'development plan' framework. The council also called on the Mapam Centre to reconsider the party's policies on expulsions not prompted by military necessity and treatment of Arabs in general with a view to curbing such things as 'racial discrimination . . . and robbery'.[46]

Taken all in all, the Yishuv's political parties had little influence on the shaping of the Yishuv's policy towards the Palestine Arab communities. The religious and liberal centre parties were weak and ineffectual. The Revisionists, considered by the majority as 'outside the fold', had a limited effect through IZL and LHI operations and through the (limited) influence of their views on sections of public opinion.

Mapai was something else. In many senses, the party was all-powerful. In 1948 its leaders became Israel's leaders. Led by Ben-Gurion, they decided on and carried out policy. But the *party institutions* consciously abdicated interest in and influence over defence and foreign policy; the unspoken feeling and motive was: 'Let our / the country's leaders get on with the job.' And the stalwarts by and large (Sprinzak was an exception) were happy with the policies and performance of the leaders—Ben-Gurion, Shertok (Sharett), Kaplan, David Remez, Golda Myerson (Meir), Levi Shkolnik (Eshkol). So there was virtually no debate within the party on policy towards the Arabs.

Not so in Mapam, where the party institutions from May 1948 onwards frankly and exhaustively discussed 'the Arab question'. The

---

[46] HHA 5.20.5 (4), protocols of the meeting of the Kibbutz Artzi Council, Nahariya, 10–12 Dec. 1948; and in HHA *Yediot Hakkibutz Ha'artzi* (Kibbutz Artzi Bulletin) of Jan. and Feb. 1949, which contain edited excerpts from the council deliberations.

party held two of the thirteen Cabinet posts and had disproportionately strong representation among the Haganah / IDF (especially Palmah) high-ranking officers. But Mapam proved unable to bring its full weight to bear on national, military, and political decision-making and operations. In large measure, this was due to the congenital split in the party between 'soft-liners' and 'hard-liners', roughly running along the Hashomer Hatza'ir–Ahdut Ha'avodah divide. The clash between the Marxist and defence-'activist' philosophies of the two groups sapped the party's resoluteness, and substantially softened or disembowelled decisions taken by party bodies on the Arab question.

This connected with the party's second major problem, the application or implementation of the party line in the field, by the 'party's' commanders in the Haganah / IDF. Ahdut Ha'avodah's general, if rarely openly enunciated, support of the 'transfer' approach to the Arab problem from the start rendered the idea of the party disciplining 'its' colonels and generals essentially absurd.

To this must be added the third paralysing factor—the common interest of all the party's kibbutzim and its two kibbutz movements in laying hands on the abandoned Arab lands in order to expand and to establish new settlements. Nor did the kibbutzim want hostile Arab communities in their vicinity.

The 'compromise' struck, as it were, was the continued enunciation of fraternal Jewish–Arab principles while in practice doing little or nothing to curb 'transfer' activities in the field. The party bodies and leaders raged or murmured against expulsions and the destruction of villages while, in the field, the party's kibbutzim and generals played a prominent part in these activities. Moreover, both kibbutz movements went ahead with the harvesting of the Arab fields, the take-over of the abandoned lands, and the establishment of new settlements. Professing support for a refugee return and for leaving aside 'surplus' tracts for returnees was, in the context, as Ahdut Ha'avodah saw things, innocuous and meaningless. But a handful of Hashomer Hatza'ir activists throughout 1948 kept up the criticism of the state's behaviour towards the Palestinians, and of their party's compromises. And a number of party stalwarts concretely assisted certain Arab communities and prevented their displacement.

By the end of 1948, even the limited, ambivalent dissent of Mapam had become largely irrelevant. Following the Kibbutz Artzi council, Mapam emerged from Israel's first general elections in January 1949 with nineteen Knesset members to Mapai's forty-six. Ben-Gurion formed a Mapai-led coalition with centre and religious parties, without Mapam. Mapai's policies towards the Arab minority (Military Government), towards the refugees (no return), and towards the Arab states (no territorial concessions) were among the reasons Mapam remained outside the coalition.

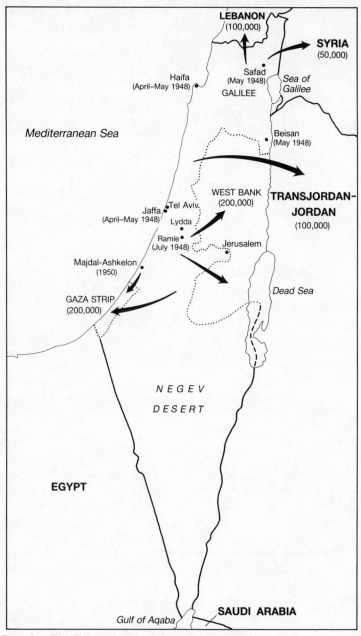

FIG. 2. The Palestinian/Arab Exodus, 1948–1950
(rough numbers: the total of Palestinian Arab refugees in 1948–50 was
600,000–760,000)

3

The Causes and Character of the Arab Exodus from Palestine: the Israel Defence Forces Intelligence Service Analysis of June 1948

Since 1948, two contradictory explanations have dominated the historical debate about the causes of the Palestinian Arab exodus. The 'traditional' Arab explanation was that the Yishuv had mounted a pre-planned, systematic campaign of expulsion already unleashed in the first months of the first Israel–Arab war. The official Jewish explanation was that the exodus had been part of a 'plot' in which the Arab leaders, inside and outside Palestine, had asked or ordered the Palestinian masses to flee their homes in Jewish-controlled territory in order to embarrass the emergent Jewish state, to justify the subsequent Arab invasion of 15 May, and to clear the ground physically, as it were, for the advance of the invading Arab armies.

The events in Palestine in 1948–9, which resulted in the Arab mass exodus, were far more complex and confused than either coherent explanation indicates. A great deal of fresh light is shed on the multiple and variegated causation of the Arab exodus in a document which has recently surfaced, entitled 'The Emigration of the Arabs of Palestine in the Period 1/12/1947–1/6/1948 (*t'nu'at ha'hagira shel arvi'yei eretz yisrael ba't'kufa 1/12/1947–1/6/1948*)'.[1]

I would like to thank the warm and efficient archivists at the Hashomer Hatza'ir Archive for helping me find material and for giving me access to the Aharon Cohen Papers before they had been finally organized and made generally available.

[1] The document is be found in the newly organized and released private papers of Aharon Cohen, the long-standing director of the Mapam (United Workers Party) Arab Department and a leading Middle East affairs expert, in the Hashomer Hatza'ir Archive (Givat Haviva, Israel), 10.95.13 (1). A notation in Cohen's hand on the cover page of the document says: 'Sent—8/7/48—Received 11/7/48'. Apparently it was sent to him by a contact in the IDF Intelligence Service or in the General Staff.

All quotations in this article are from the Intelligence Service report unless otherwise stated.

Using the term 'IDF Intelligence Service' in this context is something of a misnomer. The Haganah's Intelligence Service (the Shai), founded in the mid-1930s,

Dated 30 June 1948, it was produced by the Israel Defence Forces
Intelligence Service during the first weeks of the First Truce (11
June–9 July) of the 1948 war. The document consists of two parts,
typewritten in stencil foolscap pages: a nine-page text and a fifteen-
page appendix. The text analyses the number of refugees, the stages
of the exodus, its causes, the destinations of the refugee communities,
and the problem of their initial absorption in the host areas. The
appendix, proceeding district by district, traces—village by village
—the dates, causes, and destinations of the emigration, and the
numbers involved. The details in the appendix serve in large
measure as the basis for the statistical breakdown in the text. The
report does not state who ordered the Intelligence Service to
produce the analysis and why. It is possible that the analysis was
produced at the behest of Defence Minister and Prime Minister
David Ben-Gurion or acting IDF chief of staff and OC Operations,
General Yigael Yadin. These men, like other members of the newly
formed IDF General Staff, no doubt wanted to understand the
Palestinian exodus, which had at first surprised, indeed astonished,
the Yishuv leaders.[2]

The weeks of the First Truce gave the Intelligence Service officers
their first prolonged respite in more than six months from the
demands of daily, battle-geared operations. The major waves of
Palestinian emigration (before June 1948) had occurred in the

continued to function as the Yishuv's main intelligence service down to the summer
of 1948. At the end of May 1948 the Haganah itself became the Israel Defence
Forces, the army of the new State of Israel. But the underground army's Intelligence
Service, for bureaucratic reasons, in fact continued to exist through June, and was
only reorganized and renamed the IDF Intelligence Service at the beginning of July.
But the fact that officially the Haganah ceased to exist by the start of June makes it
incongruous to speak of a report produced by the Haganah Intelligence Service at the
end of June. (The IDF Intelligence Service, incidentally, in 1949, after a shake-up
involving the dismissal of its head, Lieutenant-Colonel Isser Be'eri, and its merger
with the already-existing IDF Operations / Intelligence Department, was renamed
the IDF Intelligence Department which, in turn, became IDF Intelligence Branch—
its current name—in 1953).

   [2] For the Yishuv's astonishment at the exodus, see, for example, the memorandum
by Israeli Foreign Minister-designate Moshe Shertok (Sharett) on his meeting in
Washington with US Secretary of State George Marshall, Dean Rusk, and other State
Department officials on 8 May 1948, in Gedalia Yogev (ed.), *Documents on the
Foreign Policy of the State of Israel, May–September 1948*, I (ISA, Government Press,
1981), 758, 760. At the meeting, Shertok referred to 'the astounding phenomenon' of
the Palestinian exodus, and said 'something quite unprecedented and unforeseen
was going on'.

preceding weeks, during the second half of April and in May, making an analysis of the phenomenon topical and relevant. Added urgency was perhaps provided by the political context. Internally, elements in the left-wing Mapam Party (The United Workers Party), a mainstay of the Israeli coalition government, began during May and June to berate Ben-Gurion and his dominant Mapai Party (Land of Israel Labour Party) openly for waging a 'war of expulsion' against the Palestinians. In the international arena, the Palestinian refugee problem moved in June to centre stage. Instrumental in pushing the refugee problem to the fore was the newly appointed UN Mediator for Palestine, Count Folke Bernadotte, who that month began his peace shuttles around the Middle East.

But, to judge from its conclusions, the Intelligence Service analysis of the exodus was hardly produced with an eye to easing the situation of the Israeli negotiators in their dealings with the Mediator, the UN in general, or the US. Rather than suggesting Israeli blamelessness in the creation of the refugee problem, the Intelligence Service assessment is written in blunt factual and analytical terms and, if anything, contains more than a hint of 'advice' as to how to precipitate further Palestinian flight by indirect methods, without having recourse to direct politically and morally embarrassing expulsion orders. 'The factor of surprise, prolonged [artillery] barrages making loud explosive sounds, [use of] loudspeakers in Arabic [to spread frightening 'black propaganda' messages], proved their great efficacy when used properly (as in Haifa particularly)', states the report. And, under the heading of 'general comments', the report adds: 'Incidentally, no attempt was made to attach fearful-sounding sirens to the wings of aircraft bombing enemy points—their effect could be great.' The comment is included in a discussion of means which might precipitate civilian flight.

This detour into advice is the only departure in the documents from straightforward analysis, whose aims, as explained in the 'general introduction', are 'to measure the dimensions of the emigration and its various stages of development, to elucidate the various factors which directly bore upon [caused] the movement of population and to indicate the destinations of the exodus'.

The Intelligence Service then gives an assessment of the number of refugees involved, allowing for a 10–15 per cent margin of error regarding the refugee population from areas inside the Jewish state as defined by the 1947 UN Partition Plan Resolution. A greater measure of inaccuracy, states the report, must be allowed for in its estimates of refugee numbers from areas lying outside the 1947 Jewish state boundaries. The facts and figures cited below, it must be emphasized, are for the period up to 1 June 1948 (except for the Jenin area, also included in the analysis, whose population fled in the last week of May and during the first week of June).

On the eve of the UN Partition Plan Resolution of 29 November 1947, according to the report, there were 219 Arab villages and four Arab, or partly Arab, towns in the areas earmarked for Jewish statehood—with a total Arab population of 342,000. By 1 June, 180 of these villages and towns had been evacuated, with 239,000 Arabs fleeing the areas of the Jewish state. A further 152,000 Arabs, from 70 villages and three towns (Jaffa, Jenin, and Acre), had fled their homes in the areas earmarked for Palestinian Arab statehood in the Partition Resolution, and from the Jerusalem area. By 1 June, therefore, according to the report, the refugee total was 391,000, give or take about 10–15 per cent. Another 103,000 Arabs (60,000 of them Negev beduin and 5,000 Haifa residents) had remained in their homes in the areas originally earmarked for Jewish statehood. (This figure excludes the Arabs who stayed on in Jaffa and Acre, towns occupied by Jewish forces but lying outside the 1947 partition boundaries of the Jewish state.)

The Intelligence Service identified four stages in the Palestinian exodus up to 1 June. Stage one, December 1947–February 1948, affected only a small number of places and involved a relatively small number of refugees, mainly from the coastal plain. The reference in the report is to the flight of much of the Arab middle class from the towns of Haifa and Jaffa.

Stage two, covering March, involved only a small number of emigrants. Emigration in general that month was in decline, but the report registers an increase in the exodus from the Jaffa and the Sea of Galilee areas.

In stage three, during April 1948, there was a 'moderate increase' on almost all fronts in the rate of emigration, according to the report. The Intelligence Service ascribes the increase to the Arab evacuation of Tiberias (18–19 April), Haifa (22 April–1 May) and the Tel-Hai (Galilee Panhandle) districts, which were a result of major Haganah offensives in those areas in the second half of April.

Stage four, in May 1948, is defined as 'the main and decisive stage in the emigration movement of the Arabs of Palestine. A psychosis of emigration began to develop, a crisis of confidence in Arab strength.' As a result, says the Intelligence Service, there was a great increase in the rate of emigration from the Tel-Hai, Gilboa, Jaffa, and Western Galilee districts, and evacuation began of the Arab villages in the Negev. May was the 'record month' of the Arab exodus, according to the report. In the predominantly Jewish coastal plain, May marked 'the final chapter', meaning that all the area's Arab inhabitants fled, except for only a few villages. It was 'the end of the job (*siyum hamelacha*)'.

Two comments are perhaps worth making about the report's analysis of the stages of the Arab exodus: (*a*) The Intelligence Service does not provide any statistical breakdown of the numbers involved in each stage, and (*b*) the description of the rate of emigration in April as 'moderate' appears questionable, in the light of the large numbers of refugees caused by the Haganah conquest of Haifa, and the major Haganah offensives in the Galilee and along the approaches to Jerusalem.

Looking to the causes of the Palestinian exodus, the core of the report, the Intelligence Service first clears the ground by dismissing the relevance of a number of factors. The report states:

One can assume that this emigration did not come as a result of economic factors—be it a serious lack of employment, food or any other economic hardship. The Arab economy [during the period up to June 1948], so long as the inhabitants stayed in their places, was not damaged in a manner which destroyed the population's capacity to subsist. The economic factor was a motive force [for emigration] only in the very earliest stages of the exodus, when the rich Arabs sought to safeguard their property and firms by getting out quickly.

According to the report, there were 'fluctuations' in the state of the Palestine-Arab economy in the cities, 'which was a factor accelerating emigration for some social strata'. But, taking the broad view, the economic factor was not 'a serious factor when speaking of the mass emigration of the Arabs of Palestine'.

The report then goes on to dismiss as precipitants of the exodus what it defines as 'pure political factors'. Political decisions and developments, in the narrow sense of the word, had 'no effect on the exodus', states the report. The Intelligence Service went on to reject specifically any linkage between the major political developments of May—the British withdrawal, the establishment of the state of Israel, the Arab declarations of intent to destroy the Jewish state and to go to war—and the mass emigration of that month. 'It must be noted here that if there were places where the political factor was a motive force in the exodus, then it was limited to the cities and, there, to a very limited social class.'

What the Intelligence Service is saying here is that Arabs did not leave the areas of the Jewish state because of opposition to the establishment of the state or political opposition to the prospect of life under Jewish rule. According to the detailed survey in the appendix, only one Arab village or community, Arab Jallad (?), in the coastal plain, fled on 15 May, because of 'the influence of the declaration of establishment of the Jewish state'.

The report then outlines what the IDF Intelligence Service regards, in June 1948, as the factors which precipitated the exodus, citing them 'in order of importance':

1. Direct, hostile Jewish [Haganah / IDF] operations against Arab settlements.
2. The effect of our [Haganah / IDF] hostile operations on nearby [Arab] settlements . . . (. . . especially—the fall of large neighbouring centres).
3. Operations of the [Jewish] dissidents [the Irgun Z'va'i Leumi and Lohamei Herut Yisrael].
4. Orders and decrees by Arab institutions and gangs [irregulars].
5. Jewish whispering operations [psychological warfare], aimed at frightening away Arab inhabitants.
6. Ultimative expulsion orders [by Jewish forces].
7. Fear of Jewish [retaliatory] reponse [following] major Arab attack on Jews.
8. The appearance of gangs [irregular Arab forces] and non-local fighters in the vicinity of a village.
9. Fear of Arab invasion and its consequences [mainly near the borders].
10. Isolated Arab villages in purely [predominantly] Jewish areas.
11. Various local factors and general fear of the future.

The Intelligence Service then gives a detailed breakdown and explanation of these factors, stressing that 'without doubt, hostile [Haganah / IDF] operations were the main cause of the movement of population'.

The wave of emigration in each district, explains the report, followed hard upon 'the increase and expansion of our [Haganah / IDF] operations in that district'. May brought a major increase in large-scale Jewish operations; so it also witnessed the widespread mass emigration of Arabs. 'The departure of the British . . . of course helped the [Arab] evacuation, but it appears that the British withdrawal freed our hands for action more than it influenced the [Arab] emigration directly.'

The Intelligence Service notes that it was not always the dimensions of a Jewish attack which counted: it was 'mainly the psychological' factors which affected the rate of emigration. The report cites 'surprise', protracted mortar barrages, and use of loudspeakers broadcasting threatening message as factors which had a strong influence in precipitating flight.

An attack on one village or town often affected its neighbours. 'The evacuation of a certain village because of an attack by us prompted in its wake many neighbouring villages [to flee]', states the report. This was especially true of the fall of large villages or towns. 'The fall of Tiberias, Safad, Samakh, Jaffa, Haifa and Acre engendered in their wake many waves of emigrants.' The psychological motive force in operation here was '*im ba'arazim nafla shalhevet*' ('If the cedars caught fire . . .', a paraphrase of 1 Kings 5: 13).

The report concludes: 'It is possible to say that at least 55 per cent of the total of the exodus was caused by our [Haganah / IDF] operations and by their influence'. To this the Intelligence Service adds the effects of the operations of the dissident Jewish organizations, 'who directly [caused] some 15 per cent . . . of the emigration'. The Intelligence Service notes that the activities of the dissidents were of especial importance in the Jaffa–Tel Aviv area, in the coastal plain to the north, and around Jerusalem. 'Elsewhere, they had no direct effect on the [Arab] evacuation.'

The Intelligence Service cites the 'special effect' of the dissident operation in Deir Yassin and of the 'abduction [at the end of March, 1948] of the five [Arab] notables at Sheikh Muwannis [north of Tel Aviv]'.

The action at Deir Yassin, especially, greatly affected the thinking of the Arab; not a little of the immediate flight during our [Haganah / IDF] attacks, especially in the central and southern areas . . . was due to this factor, which can be described as a decisive accelerating factor (*gorem mezarez mach'ri'a*).

Regarding the coastal plain, 'many of the villagers . . . began fleeing following the abduction of the notables of Sheikh Muwannis. The Arab learned that it was not enough to reach an agreement with the Haganah and that there were "other Jews" of whom to beware, and possibly to beware of more than of the Haganah, which had no control over them [that is, over the dissidents]'. The dissident organizations also played a decisive role in the evacuation of Jaffa and the villages around it, states the report. Altogether, the report states, Jewish—meaning Haganah / IDF, IZL, and LHI—military operations (comprising categories 1, 2, and 3) accounted for 70 per cent of the Arab exodus from Palestine.

Category 4: orders and commands by local Arab commanders and leaders, the Arab Higher Committee, and the Transjordan government—accounted for some '5 per cent of the villages' evacuated, according to the Intelligence Service. These orders to evacuate were given for 'strategic reasons . . . out of a desire to turn the village into a base for attack on the Jews or out of an awareness that there was no possibility of defending the village or out of a fear that the village could turn into an [anti-Arab] Fifth Column, especially if it reached an agreement with the Jews'. The latter cause was especially important in the Gilboa area (threats by the Arabs to leave directed at the Zu'abiya beduin), in the Sea of Galilee area ('Circassian villages'), in the Tel-Hai district along the Syrian border, and 'in the Jerusalem area (Arab Legion orders to evacuate a string of villages to set up bases in northern Jerusalem, and the order of the Arab Higher Committee to Issawiya [to evacuate])'.

Category 5: Jewish 'whispering' (psychological warfare) operations, usually involving 'friendly advice' by Jewish liaison officers to Arabs to quit their villages—according to the IDF Intelligence Service (which ran the liaison officers), accounted for only some 2 per cent of the exodus nation-wide. But in a number of regions, states the report, 'whispering' campaigns were of considerable importance. In the Tel-Hai district, for instance, such a campaign in April–May accounted for 18 per cent of the Arab exodus, and in the coastal plain villages, for 6 per cent. In the coastal plain and in the district, whispering operations were disorganized and unsystematic. But in the Tel-Hai district 'the operation was carried out with

predetermination, with relatively wide scope and organization'—
and so led to greater results. The operation itself was carried out,
explains the report, in the form of 'friendly advice' by Jews to their
neighbouring Arab friends.

Category 6: orders of expulsion by Jewish forces to Arab villages
—accounted (up to the start of June 1948) for some 2 per cent of the
total of villages evacuated, said the report. Such orders were
especially 'prominent' in the coastal plain, less common in the
Gilboa district, and still less in the Negev. 'Of course, the effect of
[such an] ultimatum, like the effect of "friendly advice", came after
a certain laying of the groundwork through hostile [Jewish] operations
in the area. Therefore, such [expulsion] orders are more [in the
nature of] a final motivation and propellent, than a decisive factor.'

Another 1 per cent of the emigration was caused, according to the
report, by category 7—Arab fear of Jewish retailiation after an
Arab attack on Jews. This occurred in the Western Galilee (following
the Arab attack on the Yehiam convoy), and after the attacks in
April on Kibbutz Mishmar Ha'emek (western Jezreel Valley), and
Kibbutz Gesher (Jordan Valley). According to the report, less than
2 per cent of the exodus was caused by categories 8, 9, and 10
combined. The arrival of Arab irregular forces in a village, villagers'
fears that the impending Arab invasion would turn their homes into
a battleground, and the fact of being an isolated village in a
predominantly Jewish area all had little effect on the villagers.

The report names two further direct causes of flight: 'general fear'
and 'local factors'. General fear, which 'had a great influence and
role in the exodus', accounted for some 10 per cent of the refugees.
In this context the report mentions the initial waves of emigration at
the start of the hostilities, caused 'at first glance, by no special
reason'. These were rooted in a 'general fear' resulting primarily
from 'the crisis in confidence in Arab strength'.

The Intelligence Service thus places this 'crisis of confidence' in
the Arab power to fight and withstand or defeat Jewish arms as 'the
third most important factor, after our own [i.e., Haganah / IDF]
operations and those of the dissidents', in the Arab exodus. The
report states that 8–9 per cent of the exodus was caused by 'local
factors', such as the breakdown in specific localities of Arab–Jewish
peace negotiations and the Arabs' 'inability to adjust to certain real
situations'.

Following this statistical breakdown, the report offers some 'general comments' identifying some direct and indirect contributory factors which hastened, precipitated, or increased waves of emigration in various areas at different times. First and foremost, the report refers to a 'psychosis of evacuation' which gripped some Arab communities during the hostilities, 'increasing the rate of evacuation'. It appeared, stated the report, 'like a contagious disease'. As an example, the Intelligence Service cites the case of Acre, which fell to Haganah forces on 17 May. There 'it is possible to assume . . . that the massive arrival on the scene [a fortnight before] of the refugees from Haifa, who planted in the hearts of Acre's inhabitants a psychosis of evacuation . . . had a decisive influence'. Thus, 'light attacks' and 'nudges' by the Jewish forces around Haifa had the effect of precipitating flight in a population already affected by 'evacuation psychosis'. The appearance of typhus also prompted flight. 'More than the disease itself', states the report, 'the panic created by the rumours of the spread of the epidemic was a factor prompting evacuation.' The report points out that where there was a 'strong Arab military force' the villagers did not evacuate 'readily', and 'only a direct and serious operation [by the Jewish forces] brought about the destruction of this [military] force, bringing flight [of the civilian population] in its wake'.

At the start of the evacuation 'the Arab institutions attempted to struggle against the phenomenon of flight and evacuation, and to curb the waves of emigration'. The Arab Higher Committee decided to impose restrictions, and issued threats, punishments, and propaganda in the radio and press to curb emigration. The committee also tried to mobilize the governments in the neighbouring Arab states to assist in this; there was a coincidence of interests. 'Especially, they tried to prevent the exodus of youngsters of military age', states the report. 'But all these actions completely failed because no positive action was taken which could have curbed the factors pushing towards emigration.' The sole upshot of these efforts was corruption and bribery, whereby officials began selling permits to would-be emigrants wishing to leave Palestine or to enter other countries. But this arrangement, states the report, broke down once emigration turned into a mass movement.

The penultimate section of the report deals with the destinations of the refugee communities. The authors note certain patterns. For instance, city-dwellers of rural origin often returned to their ancestral villages, as did Jaffa residents who had originated in Faluja, for example. Similarly, according to the Intelligence Service, city-dwellers who had come, or whose fathers had come, from neighbouring countries tended during the hostilities to return to those countries. Thus many Haifa residents fled directly to Lebanon and Syria. In general, the report points out, urban dwellers, including the rich, had fled directly to their final destinations whereas rural refugees tended to 'hop' through a number of way-stations before reaching their final point of rest.

Often, villagers fled at first from isolated rural sites to neighbouring Arab towns or cities. Then, when the town fell to Jewish forces, they moved on. The report cites the case of the inhabitants of Beit Sussin, in the south, who first fled to Mughar, then Yibna, and then Isdud, before coming to rest in Gaza. By and large, the refugees from the villages in the first instance moved only to nearby sites. This, according to the report, caused the Haganah problems; the Jewish units, which did not have sufficient troops to garrison every captured village, faced the prospect that the villagers might attempt to return to their homes. 'More than once,' states the report, '[Haganah / IDF units were forced] to expel inhabitants [after they had returned to their homes].'

Without giving a numerical breakdown, the report states that the wealthier Arabs mainly emigrated directly to Arab states (meaning, primarily, to Beirut and Cairo); the poorer Arabs of the northern border areas fled to Syria and Lebanon: and the inhabitants of the south and Jaffa, and some Jerusalemites and Haifa residents, moved to Egyptian-held territory (meaning mainly the Gaza strip). Most of the emigrants to Transjordanian-held territory came from the Sea of Galilee area, the Jezreel valley, the Gilboa area, Acre, Jaffa, and the Jerusalem area.

The report ends with a look at the manner in which the refugees (by June 1948) had been absorbed in the host countries or areas. The wealthier Arabs, by and large, had no absorption problems. But most of the emigrants were poor, most had left without the bulk of their belongings, and this had led to 'severe absorption problems', says the report. This had prompted the governments of the host countries to try to persuade the refugees to go back and to put pressure, especially on able-bodied males, 'to return to the front'.

Transjordanian radio from Jerusalem, for example, in May broadcast lectures to the Palestinians to go back and lend a hand in the war effort, maintaining that 'there was no danger to the lives of those returning'.

Some Israelis feared that the embittered refugees might be turned into soldiers who would return to fight against Israel. The Intelligence Service analysis dismissed this danger: 'The Arab emigrant did not turn into a fighter, his only interest now is in collecting money [philanthropy]. He has resigned himself to the lowest form of life, preferring it to mobilizing for battle.' In conclusion, the report states that the refugees were a burden which would continue to grow and weigh upon the Arab states, especially as 'no serious and comprehensive organized step was being taken by the Arab states in order to solve the problem'.

How accurate is the information conveyed in this document? How sound is its analysis of the causes of the Palestinian exodus up to June 1948? What is its significance in relation to the traditional perceptions of the character and causes of that exodus?

In theory at least, the IDF Intelligence Service—Israel's main intelligence service in June 1948—was very well placed to collect and analyse data about the Palestinian exodus. The officer or officers who produced this report had access to the reports of Israeli agents and Arab informants in the various Arab localities, to the signals and reports of the Haganah / IDF unit intelligence officers (one at least was attached to every battalion and brigade) and, probably, to signal traffic and reports of the various unit commanders and front commanders around the country.

It is also possible that the authors of the report were supplied, at their request, with special reports by units' intelligence officers and perhaps unit commanders as well detailing each unit's history of conquest and treatment of Arab settlements. The respite provided by the first weeks of the First Truce would have made possible the writing of such reports. An indirect indication that such reports were indeed produced and, at least in part, served as the basis of this analysis is afforded by the absence of one of the two appendices which, according to the table of contents printed on the covering page of the document, were to have accompanied the text—'appendix 1' giving 'regional surveys analysing the problems of emigration in each and every district'. Presumably, these surveys were to have been written by unit, front (*hazit*), or district (*nafa*) intelligence officers. Either some of them were not delivered or those delivered were regarded as inadequate for reproduction along with the text and the originally entitled 'appendix 2', which details the exodus from each village, by district, around the country. (Appendix 2, in fact, was included, retitled 'appendix 1'.)

In the end, the authors apparently decided that the analysis, buttressed by the village-by-village appendix, was sufficient, and the regional analyses at first contemplated were left out (though sallies into regional analysis are to be found interspersed unsystematically throughout the text).

Real-time signal traffic and post-operational reporting by and large were accurate in the Haganah / IDF in 1948. But until mid-May—covering almost the whole period dealt with in the report— the Haganah was an underground force, and did not produce or store the kind of comprehensive documentation about its operations that a good regular army would have done. Much of the reporting up the chain of command and orders handed down the chain of command in the first months of 1948 were necessarily oral, and much of the signal traffic was never recorded or was subsequently lost. Hence, in producing this analysis of the exodus, the Intelligence Service perforce had to rely to some extent on the memories of commanders and local intelligence officers rather than on contemporaneous chronicling.

Moreover, the dissident organizations—the IZL and LHI, to whose operations the report attributes some 15 per cent of the exodus —produced even less written material than the Haganah and, if it existed in June, this was never made available to the Haganah and IDF. (The Haganah regarded the IZL and LHI as hostile organizations; June, indeed, marked the high point of the conflict between the groups, with the IDF killing a number of IZL combatants and sinking off Tel Aviv a ship bearing arms for the dissident force.)

Lastly, a number of operations by local Haganah units and by Jewish settlements against neighbouring Arab communities were carried out with Haganah National Staff command, authorization, or approval, and were never accurately reported upon to the National Staff *ex post facto*.

The reservations about sourcing aside, there is no reason to

cast doubt on the integrity of the IDF Intelligence Service in the
production of this analysis. The analysis was produced almost
certainly only for internal, IDF top brass consumption. (No copy of
it has surfaced in any collection of private papers of the 1948 Israeli
Cabinet ministers; nor, save for the copy used in this paper, in any
civilian archive. Nor was its existence or content ever referred to in
any recorded political party debate.)[3] The authors of the report
would have certainly been conscious of their 'consumer public', and
aware that many of the consumers were highly familiar with parts of
the subject matter of the analysis. On the other hand, the authors
will not always have been familiar enough with given incidents to
catch all errors or distortions in the reports of local commanders and
intelligence officers. So, while the details of the report and its
analysis by and large conform with the facts as recorded in other
sources from the period, a degree of analytical inaccuracy and
factual imprecision and error is none the less evident. This point is
worth elucidating before going on to weigh up the significance of the
document.

The village-by-village survey in the appendix lists 14 villages
evacuated as a result of Haganah or IDF orders or ultimatums.[4] In
peacetime these villages together had a population of some 20,000.

[3] The report was received by Cohen, a member of the Mapam Centre and of the
party's Political Committee, on 11 July 1948. These party institutions, through
the summer, very frankly and thoroughly debated the Yishuv's policy towards
the Palestinian Arabs, covering such subjects as expulsions, the possibility of a
refugee return, etc. Yet neither Cohen nor anyone else ever referred in these
recorded debates to the IDF Intelligence Service analysis. One possible explanation
is that the report, being contemporary and on a sensitive subject, was regarded by
Cohen as something too 'hot' to use or refer to openly. Mapam, Cohen, and his
contact might have been in deep trouble. Ironically, only a few years later Cohen
landed in prison after being convicted of unauthorized contacts with Soviet agents.
[4] The villages named are Ad Dumeira (population 620—evacuated 10 Apr. 1948);
Miska (population 650—evacuated 15 Apr. 1948); Khirbet as Sarkas (evacuated 15
Apr. 1948); Arab an Nufei'at (population 910—evacuated 10 Apr. 1948); Khirbet
Azzun (population 994—evacuated 3 Apr. 1948); and Arab al Foqara (population
340—evacuated 10 Apr. 1948), all in the coastal plain; Dana (population 400—
evacuated 28 May 1948), in the Gilboa district; and Zarnuga (population 2,600—
evacuated 27 May 1948); Yibna (population 5,920—evacuated 4 June 1948); Huj
(population 800—evacuated 28 May 1948); Arab Rubin (population 1,550—evacuated
1 June 1948); Kaukaba (population 1,870—evacuated 17 May 1948); Sumsum
(population 1,200—evacuated 12 May 1948); and Najd (population 600—evacuated
12 May 1948), all in the south. It is worth noting that the expulsions detailed in the
appendix are almost all part of two 'series'—one in the northern coastal plain on 10–
15 April 1948 and the other in the northern Negev approaches on 27–8 May 1948.

Yet in the analysis of the causes of the exodus, the report speaks of only 2 per cent 'of the villages' (out of a total of 250 evacuated) as leaving because of Haganah / IDF expulsion orders. Fourteen out of 250 represents more like 5 per cent.

Moreover, the report leaves a large, poorly demarcated grey area between outright expulsion by Jewish order and evacuation of Arab villages in the course of Haganah / IDF 'military operations' (which are said to account for 55 per cent of the exodus).

Some of the villages said to have been evacuated because of 'military operations' (and presumably included in that 55 per cent), are seen in the detailed breakdown in the appendix to have been de-populated in a somewhat less straightforward manner. For example, the 710-strong population of Khirbet Lid (al-Awadim), near Afula, in the Jezreel Valley, is said in the appendix to have left because of 'the influence of [the nearby battle of] Mishmar Ha'emek' in April 1948. But in the subsequent 'comment', the appendix also states: 'They tried to return. And were expelled.' Khirbet Lid was pre-sumably not included under the expulsion category.

Nor was Fajja, a large village next to Petah Tikva. Part of the population left after the IZL attack on 17 March. The final evacuation on 15 May took place, according to the appendix, because of 'pressure by us [and] a whispering [that is, psychological warfare] campaign'. Presumably Fajja was listed among the 2 per cent of evacuations caused by psychological warfare; but, given the reference to 'pressure' by the Haganah, it could also have been included perhaps in the expulsion category (which it presumably was not).

Nor was Al Khalisa, the site of present-day Kiryat Shmona, in the Galilee Panhandle. The village, with a population of 1,840, is said to have been evacuated on 11 May because of 'the fall of Safad', a major Arab centre to the south. But according to the appendix, that was not all. 'They wanted [to reach] an agreement with us. They were turned down. [So] they fled', states the report. Presumably, Khalisa was included under the 'local factors' category rather than under the expulsion category. As in Al Khalisa, so in As Salihiya, a village of 1,520 a few kilometres to the south. 'They wanted to negotiate—we did not show up', states the report. The villgers fled Palestine on 25 May.

In general, the situation on the ground made it impossible in many cases to draw a clear distinction between a Haganah / IDF or

IZL 'military operation' which ended in villagers fleeing their homes and 'expulsion orders', which had the same effect. In some 'military operations', such as the Haganah conquest of the Arab parts of Haifa, the Jewish troops by and large had no clear intention of provoking an Arab exodus and their military strategy was not calculated to produce such an outcome. In other military operations, such as the IZL attack on Jaffa, and probably the Haganah offensive in Western Galilee in May 1948, the flight of the Arab inhabitants was clearly desired and deliberately provoked by the attacking troops. The IZL / LHI attack on Deir Yassin near Jerusalem on 9 April ended not only in a massacre but also in the expulsion by the conquering unit of the surviving Arab villagers. (The Intelligence Service report catagorizes the flight of the Deir Yassin inhabitants as a result of a dissident operation rather than under the heading of expulsion.)

While the report was not produced with any propagandizing intention in mind, its authors seem to have exhibited a perhaps understandable tendency to minimize the role direct expulsion orders played in bringing about part of the Palestinian exodus. The proportion of villages expelled is computed incorrectly and a large grey area of 'semi-expulsions' is included under the category of flight due to 'military operations' or some other 'non-expulsion' category.

Moreover, the report also includes a number of factual errors and omissions in this context; presumably these were the result of misinformation in the reports by local unit commanders and field intelligence officers. For instance, part of the population of the Arab town of Beisan (Beit Shean) is said to have fled on 1 May because of 'fear and the influence of [the fall of Arab] Haifa'. The remainder of the population, according to the appendix, is said to have left on 12 May as a result of the Haganah 'conquest [of the town]. Fear. The influence of Haifa.' But this is not completely accurate. Hundreds of the town's residents stayed on after the conquest, and were expelled only days later—some to Nazareth, others across the Jordan River—at Haganah command.[5]

[5] See Binyamin Etzioni (ed.), *Ilan Va'shelah*, an account of the Golani Brigade's operations in 1948 produced by the unit's soldiers, (IDF Publications (Ma'arachot), Tel Aviv, 1950?), 146, See also Central Zionist Archives S53–437 (the Eliezer Granovsky Papers), Yosef Weitz to Eliezer Granovsky, 25 May 1948; and Yosef Weitz, *Diaries* (Massada, Tel Aviv, 1965) iii. 301–2, entry for 13 June 1948.

The small village of 140 tenant farmers of Qira wa Qamun near Yoqne'am, on the western edge of the Jezreel Valley, was evacuated in March by its inhabitants after they received 'friendly advice' from the local Haganah intelligence officer at Yoqne'am, Yehuda Burstein.[6] But the report gives the reason for the Qira evacuation as 'fear and the influence of the attacks in the area'—not really the same thing.

More inexplicable is the omission altogether from the appendix of the fate of a string of Western Galilee villages—Az Zib, Manshiya, As Sumeiriya, Al Bassa and others—all evacuated during or before the Haganah's Operation Ben-Ami in mid-May. It is quite possible that the Haganah commander in Western Galilee or the relevant intelligence officers simply failed to submit to the Intelligence Service a report on the Arab exodus from their area.

The report's treatment of villages evacuated as a result of edicts or orders by Arab authorities, political or military, is also worth examining. Altogether, 21 villages out of the 250 are listed in the appendix as having been evacuated or partially evacuated as a result of Arab command, be it by the Arab Legion, the Arab Higher Committee, or other Arab bodies.[7] The figure is higher than the '5 per cent' cited in the report's analysis as having fled because of Arab commands. Here too the report omitted or ignored material instancing Arab advice or orders to communities to partially or completely evacuate their settlements. For example, the 'defence and security section' of the (Arab) National Committee in Jerusalem, basing itself on instructions from the Arab Higher Committee issued on 8 March, in mid-April ordered the national committees in the Sheikh Jarrah, Wadi Joz, Sa'ad wa Sa'id, Musrara, and Katamon

---

[6] Interview with Eliezer Be'eri (Bauer), Kibbutz Hazore'a., April 1984.

[7] The report lists the following villages as evacuated, or partly evacuated, at higher Arab command; Shu'fat, Beit Hanina, Al Jib, Judeira, Beit Nabala, and Rafat (total population 4,000–5,000—all on Arab Legion orders) all on 13 May 1948; Issawiya (population 780—evacuated at Arab Higher Committee command on 30 Mar. 1948); Ar Ruweis, on 24 Apr. 1948; Ad Dahi, Nein, Tamra, Kafr Misr, At Tira, Taiyiba, and Na'ura, all in the Gilboa district and evacuated on 20 May 1948, after threats from Arab irregulars; and, in the Sea of Galilee area, Adasiya (evacuated on 15 May 1948 at Transjordanian command); and Sirin, Ulam, Hadatha, and Ma'adhar (all in the Galilee, evacuated at the command of the Arab Higher Committee on 6 Apr. 1948). Within months, the populations of Shu'fat, Beit Hanina, Al Jib, Judeira, Issawiya, and Tamra had returned to their homes.

quarters of Jerusalem to order the women, children, and the old in
their areas to leave their homes and move to areas 'far away from
the dangers. Any opposition to this order . . . is an obstacle to the
holy war . . . and will hamper the operations of the fighters in these
districts.'[8]

In early May, units of the Arab Legion entered the town of
Beisan and reportedly ordered the evacuation of all women and
children.[9] At about the same time, the Arab Liberation Army was
reported to have ordered the villagers in Fureidis, south of Haifa, to
'evacuate the women and children from the village and to make
ready to evacuate the village completely'.[10] In general, the IDF
Intelligence Service report fails to stress the importance in the Arab
exodus of the early departure in many cases from the villages of the
women and children. This tended to sap the morale of the menfolk
who were left behind to guard the homes and fields, contributing to
the final evacuation of villages. Such two-tier evacuations—women
and children first, the men following weeks later—occurred in
Qumiya in the Jezreel Valley, among the Ghawarina beduin in
Haifa Bay, and in various other places.

What then is the significance of the IDF Intelligence Service report
in understanding the Palestinian exodus of 1948? To begin with, it
thoroughly undermines the traditional official Israeli 'explanation'
of a mass flight ordered or 'invited' by the Arab leadership for
political–strategic reasons. Quite clearly, according to the report,
Arab orders to evacuate villages were restricted to a number of
areas, were guided by local strategic considerations, and affected no
more than 10 per cent of the Palestinian refugee population. (About
half of the villages evacuated because of Arab command, those in
the Jerusalem and lower Galilee areas, were in fact subsequently
repopulated by their original inhabitants once circumstances had
changed.)

The report makes no mention of any blanket order issued over
Arab radio stations or through other means, to the Palestinians to
evacuate their homes and villages. Had such an order been issued, it

---

[8] ISA, FM 2570/11, announcement by the National Committee of Jerusalem, 22
Apr. 1948.
[9] CZA, A246–13 (the manuscript of the Weitz Diaries), entry for 4 May 1948.
[10] Private information.

would without doubt have been mentioned or cited in this document; the Haganah Intelligence Service and it successsor, the IDF Intelligence Service closely monitored Arab radio transmissions and the Arabic press.

Indeed, the Intelligence Service report in its main thrust seems to go still further in undermining the official Israeli historiography. For not only is the 'Arab orders' explanation seen to be limited in the numbers it affected and extremely restricted geographically; but the report goes out of its way to stress that the exodus was contrary to the political–strategic desires of both the Arab Higher Committee and the governments of the neighbouring Arab states. These, according to the report, struggled against the exodus—threatening, cajoling, imposing punishments, all to no avail. There was no stemming the panic-borne tide. (To this, a caveat must be attached. The report does not record or analyse the *dates* of the official Arab efforts to stem the exodus. The dates may be significant. Whereas there is evidence of a large number of Arab attempts to stop the exodus during December 1947 and during the first months of 1948 and in early May 1948, there is far less material of this sort relating to April and mid- and late May 1948, when the flight reached its peak.) But neither does the Intelligence Service report uphold the traditional Arab explanation of the exodus—that the Jews with premeditation, in centralized fashion, and systematically had waged a campaign aimed at the wholesale expulsion of the native Palestinian population.

The exodus was certainly viewed favourably by the bulk of the Yishuv's leadership; it had solved the embryonic Jewish state's chief and agonizing political–strategic problem, the existence in it of a very large actively or potentially hostile Arab minority. A tone of satisfaction with the exodus does indeed pervade the report; but from it emerges a very definite impression that the depopulation of the villages and towns was an unexpected outcome of operations the purpose of which was wholly or primarily the conquest of military positions and strategic sites in the course of a life-and-death struggle. Jewish military operations indeed accounted for 70 per cent of the Arab exodus; but the depopulation of the villages in most cases was an incidental, if favourably regarded, side-effect of these operations, not their aim. Had the population of the villages and towns remained *in situ* during and after the Jewish attack and conquest, the Haganah

/IDF and IZL would have been faced at each site with the successive dilemma: to expel or not to expel. As it was, the population, by taking to its heels at the first whiff of grapeshot, usually solved this possible problem. The report's estimate of the proportion of villages depopulated by calculated, direct Jewish expulsion orders is none the less somewhat low. For the period up to 1 June 1948, something around 5 per cent seems closer to the mark than the 2 per cent cited. Even after adding to this the villagers 'nudged' into flight by deliberate military pressures and psychological assault, one is still left with only a small proportion of the exodus accounted for in this manner.

One must again emphasize that the report and its significance pertain only up to 1 June 1948, by which time some 300,000–400,000 Palestinians had left their homes. A similar number was to leave the Jewish-held areas in the remaining months of the war. The circumstances of the second half of the exodus—during the IDF conquest of Lydda and Ramle, and the central Galilee in July, the northern Negev in October–November, and the northern Galilee in October—are a different story. But for an understanding of the Palestinian exodus until 1 June, one must, according to IDF Intelligence, reach mainly for the vast middle ground between pre-planned, outright IDF expulsion and Arab-engineered, Machiavellian flight. There, amid the frightening, threatening boom of guns, the loss of confidence in Arab might, the flight of relatives and friends, the abandonment of nearby towns, and a general, vast fear of the uncharted future, one will find the bulk of the pre-June Palestinian refugees.

4

Yosef Weitz and the Transfer Committees, 1948–1949

I met Ben-Gurion . . . [He said:] 'The Arabs of the Land of Israel, they have but one function left—to run away'. With that he got up and ended the conversation.

(Ezra Danin, special adviser on Arab affairs at the Foreign Ministry (Tel Aviv), to Elias Sasson (Paris), director of the Middle East Affairs Department, Foreign Ministry, 24 October 1948.)[1]

There is reason to assume, that what is being done . . . [is being done] out of certain political aims and not only out of military necessity . . . In fact, what is called a 'transfer' of the Arabs out of the area of the Jewish state is what is being carried out . . .

(From 'Our Arab Policy During the War', a memorandum to the Political Committee of Mapam by Aharon Cohen, director of Mapam's Arab Department, 10 May 1948.)[2]

The migration of the Arabs of the Land of Israel was not caused by persecution, violence, expulsion . . . [it was] a tactic of war on the part of the Arabs . . .

(The Weitz Committee Report, November 1948).[3]

As the Yishuv welcomed the 29 November 1947 UN General Assembly Partition Resolution on Palestine, its leaders grimly pondered the future prospect of war and the emergent Jewish state's major problem: of the 1.1 million or so people included in the

[1] ISA, FM 2570/11, E. Danin (Tel Aviv) to E. Sasson (Paris), 24 Oct. 1948.
[2] HHA-ACP, 10.95.10 (4).
[3] ISA, FM 2445/3, the report of the Transfer Committee (Yosef Weitz, Ezra Danin, Zalman Lifshitz), entitled 'Regarding the Solution to the Arab Refugees', Nov. 1948.

areas earmarked for Jewish independence, close to 45 per cent were Arabs. Was there a solution to 'the Arab problem', as it was commonly referred to in Yishuv leadership circles?

Within weeks, hostilities between Arab and Jew engulfed large areas of the country; within two to three months, tens of thousands of the land's Arab inhabitants, mostly from the urban middle class of Jerusalem, Jaffa, and Haifa, departed for Beirut, Cairo, and the Hebron Hills and the area then known as the 'Triangle' (encompassing Nablus, Jenin, and Tulkarm).

The steady Arab exodus took the Yishuv's leaders by surprise; for most, a pleasant surprise. With his eye for history, the Yishuv's leader, Jewish Agency chairman David Ben-Gurion, quickly recognized the exodus as the possible start of a thoroughgoing demographic-political-economic revolution which could solve the 'Arab problem'. Perhaps most or all the country's Arabs would go. 'What happened in Jerusalem and . . . in Haifa, could happen in large parts of the country if we stand fast . . . It is possible that there will be very great changes in the country in the [coming] six or eight or ten months of this war . . . Certainly there will be great changes in the composition of the population of the country . . .' Ben-Gurion told the Mapai (Land of Israel Labour Party) Council on 7 February 1948. He was speaking after a brief visit to western Jerusalem, which had become '100 per cent Jewish . . . In many Arab districts in the west one sees not a single Arab. I do not believe this will change', he said.[4]

A similar feeling that a giant, historic transformation was under way also gripped Yosef Weitz (1890–1972), director of the Jewish National Fund's Lands Department who, since the 1930s, was responsible for land acquisition (mostly from Arabs) by the Yishuv. The departing Arabs, he was convinced, would not return, and events were leading to 'a complete territorial revolution . . . The State [of Israel] is destined to expropriate . . . their land.'[5]

---

[4] Labour Party Archives (Beit Berl), 22/35, protocol of the Mapai Council meeting 7 Feb. 1948. A slightly different version is given in David Ben-Gurion, *Yoman Hamilhama-Tashah* [the war diaries, 1948–9] (Tel Aviv, 1982) i. 210–11, entry for 6–7 Feb. 1948. (The Ben-Gurion diaries henceforward will be referred to as DBG.)

[5] Yosef Weitz, *Diaries*, 5 vols., iii, (Tel Aviv, 1965), 288, entry for 20 May 1948. (The Weitz Diaries will henceforth be referred to as Weitz.)

Weitz early on realized that the state of anarchy created by the hostilities and the steady breakdown of British Mandate rule could and should be exploited if the 'Arab problem' was to be solved and the Jewish state established on solid foundations. He was excellently placed to do this exploiting and to egg others on to do the same.

By 1949, the Yishuv's main pre-state 'National Institutions'—the Jewish Agency, the JNF, the National Committee (*ha-Va'ad ha-leumi*) and so forth—had lost many or all of their powers, giving way to the new institutions of state, primarily the government ministries, the Cabinet, and the Knesset. But through 1948, while the ministries were still being established, the National Institutions retained major powers. As one of the National Institutions' chief executives, Weitz in 1948 still sat astride the crossroads of power in the new state.

As the JNF executive in charge of land acquisition and its allocation to Jewish settlements, as the JNF representative on the Committee of Directorates of the National Institutions (*Va'ad ha-Hanhalot shel Hamosdot ha-Leumim*) and on the Settlement Committee of the National Institutions, as chairman of the Negev Committee (the civilian government of the Jewish settlements in the southern desert region) and as a member of the three-man Arab Affairs Committee of the National Institutions, Weitz was well placed to shape and influence decision-making regarding the Arab population on the national level and to oversee the implementation of policy on the local level. The JNF's branches and personnel around the country were at his disposal.

To old and new Jewish settlements, Weitz was a powerful patron figure, the dispenser of much-needed tracts of land. For the Yishuv's political leaders, he was an expert on territorial, settlement, and Arab matters. Through 1948 he had ready access to key cabinet ministers, such as Foreign Minister Moshe Shertok (later Sharrett) and Finance Minister Eliezer Kaplan—and often, he met with Ben-Gurion, who from mid-May was Prime Minister and Defence Minister. Weitz's connections also encompassed the Yishuv's military brass, especially on the level of district, area, and battalion commanders, most of whom came from the kibbutzim with which Weitz had continuous dealings. His expertise on land matters and Arabs—stemming from years of organizing land deals and regional development projects—was the basis of his standing among the

Yishuv's intelligence apparatus, the Haganah's Sherut Yediot (the Shai), and its successor organization, the IDF Intelligence Service.

EVICTING TENANT-FARMERS AND BEDUIN

Following the start of Arab–Jewish hostilities at the end of 1947, Weitz was among the first to apppreciate and act upon the potential for land acquisition inherent in the state of anarchy which had descended upon Palestine as rival militias traded blows under the eyes of British regiments preoccupied with organizing their own safe withdrawal from a land that had become ungovernable. His attention in early 1948 initially focused on the Arab tenant-farmer communities which 'squatted' on JNF-owned land. Absentee Arab landowners had previously sold the land to the JNF but the tenants in some areas had refused JNF compensation offers and refused to move off, or accepted compensation and then stayed put. In some cases during the 1940s, the JNF, assisted by local Haganah forces, had destroyed such squatters' houses and fields in an effort to drive them off. In the hilly Ramot–Menashe area, south-east of Haifa, three such tenant communities—Ju'ara (with 14 families), Daliyat ar Ruha (with 53 families) and Qira wa Qamun (with 42 families)— had been compensated and in the main evicted in the mid-1940s. But some of the tenant families had stayed on and others had returned to Daliyat ar Ruha and Qira, rebuilding their houses and renewing cultivation of the fields. The British Mandate authorities barred the Yishuv from using force to evict them.[6]

In early 1948 Weitz saw his chance. Travelling north on 11 January, Weitz met in Haifa with officials of the JNF's Northern District office to discuss the situation in Ramot–Menashe, an area strategically, economically, and politically vital to the Yishuv as it linked the Jewish population centre in the coastal plain with the Haganah and kibbutz centre in the Jezreel valley. Despite its predominantly Arab population, the area was included by the November 1947 UN Partition Plan in the territory of the prospective Jewish state. 'Was not now the time to be rid [of the tenant-farmers]? Why continue to leave in our midst these thorns which represent a

---

[6] Central Zionist Archives, (CZA) S-25/10682, a memorandum: 'On the Question of Eviction of Arabs' by Yosef Weitz, 20 Mar. 1946.

potential danger . . .? Our people are [now] weighing up . . . the matter,' Weitz noted in his diary after the meeting.[7]

The following day Weitz journeyed to Yoqne'am, where he met the local Haganah intelligence officer, Yehuda Burstein. The two 'discussed . . . the question of the eviction of tenant-farmers from Yoqne'am and Daliyat [ar Ruha] with the methods now acceptable. The matter has been left in the hands of the defence people [that is, the Haganah] and during the afternoon I talked with the [Haganah] deputy commander in the district.'[8] But the Haganah as yet had formulated no policy regarding the Arab population of Palestine; local commanders had no authority to evict Arabs. The matter hung fire. But Weitz persisted. On 22 February he participated in what he defined as 'a final discussion' about 'the evacuation of our lands in Yoqne'am, Daliyat and [Qira wa] Qamun, and the means that must be employed [to achieve this]'.[9] A few weeks later, Burstein ordered the tenant-farmers of Qira wa Qamun off the land, on pain of attack by the Haganah. They left. So did those living in Arab Yoqne'am and most of those at Daliyat ar Ruha, possibly after similar intimidation by Burstein or in a domino reaction, based on fear, following the exodus from Qira wa Qamun.[10]

In late March, Weitz, on another visit to Haifa, noted that the tenant-farmers at Yoqne'am had departed. The 'problem' of what to do with their fields and shacks remained. It was decided, in a discussion with JNF officials, 'to take everything into our possession . . . to dismantle the shacks . . . and to pay the tenant-farmers compensation . . . .'.[11]

Daliyat, however, remained a problem. Some 12 families had remained *in situ*. It was decided to pay them 'what was due in compensation, on condition they leave the site. If they don't agree [to go]—we will force them to do so. They cannot be left in our midst because of the danger that alien gangs [of Arab irregulars] will

[7] Weitz, iii, p. 226, entry for 11 Jan. 1948.

[8] CZA, A-246/12, p. 2290, entry for 12 Jan. 1948. Yosef Weitz's diaries were published in a somewhat abridged and edited form by Weitz in 1965. The originals of the diaries, in manuscript notebooks are deposited in the CZA, in file A-246. The differences between the original manuscripts and the published diaries are important especially with regard to the subject dealt with in this essay.

[9] CZA, A-246/13, pp. 2,315–16, entry for 22 Feb. 1948.

[10] Interview with Eliezer Be'eri, Kibbutz Hazore'a, Apr. 1984, for information on Burstein and the intimidation of Qira.

[11] Weitz, iii, pp. 256–7, entry for 26 Mar. 1948.

infiltrate into their midst and attack our settlements.'[12] The problem of Daliyat ar Ruha remained unresolved. On 2 April Weitz raised the matter with one of Ben-Gurion's senior aides, Levi Shkolnik (later Eshkol, who became Ben-Gurion's successor as Prime Minister in 1963). Weitz complained that the Haganah 'did not want to act there'. The discussion more generally covered 'the eviction of Arabs' from Jewish-owned lands. Shkolnik said that he knew nothing about the Daliyat problem, and Weitz concluded: 'It appears that I will personally have to deal with all [matters] pertaining to JNF tracts.'[13] In mid-April, the last tenant farmers left Daliyat ar Ruha, and on 29 July 1948 a new Israeli settlement, Kibbutz Ramot-Menashe, was established near the site.

While in Haifa in the early months of 1948, Weitz also took a keen interest in another problem area, the Beit Shean (Beisan) Valley, where the JNF had long sought to buy out Arab landholdings to set up new Jewish settlements and to establish a continuous concentration of Jewish habitation and cultivation where full-scale regional planning would be possible. (Under the Mandate, Jewish rural regional planning, in which Weitz figured largely, was almost impossible because of the intermixed populations and landholdings in every district.) In late February Weitz noted that the semi-nomadic beduins of the valley—Al Bawati, Al Ghazawiya, As Safa, Az Zarraa, Al Khuneizir, Al Hamra, Al Arida, Umm Ajra, and As Sakhina—were beginning to decamp and move across the Jordan River to Transjordan. 'Possibly now is the time to activate our original plan and to transfer them there', Weitz wrote.[14]

But the situation in the valley, part of the future Jewish state according to the partition plan, was complex. Arab cultivators and roving beduins lived on Jewish-owned land, on Arab lands, and on state-owned lands, which until the British evacuation in May were the property of the Mandate Government. Some of the state lands were formally leased to Arabs until long after the termination of the Mandate. At Al Ashrafiya, some Arab residents had pulled out. The question, for Weitz and his Haifa JNF staff, was whether 'to seize [the Ashrafiya land] now. My opinion was negative', wrote Weitz. 'We have no interest in entering into conflict with the

---

[12] Weitz, iii, p. 257, entry for 26 Mar. 1948.
[13] Ibid., p. 261, entry for 2 Apr. 1948.
[14] Ibid., pp. 239–40, entry for 20 Feb. 1948.

[British] government over the land. What must be done: To make sure the beduins don't occupy the tracts and to be ready, come May . . . to take over the land ourselves.'[15]

But the beduins began to trickle back into the valley at the end of March and beginning of April. Weitz took up the matter with his Haifa staff and the local Beit Shean Haganah commanders on 26 March. Moshe ('Musa') Goldenberg, Haganah area commander (*maaz*) of Kibbutz Beit Alfa, and a representative of Kibbutz Maoz, 'demanded' the immediate establishment of another Jewish settlement near Kibbutz Kfar Ruppin 'as a means of liberating our land and preventing the return of the beduins who had fled to Transjordan . . . At my insistence', wrote Weitz, 'it was agreed that the [Jewish] settlements in [the] Beit Shean [Valley] would begin to receive the lands and cultivate them. As to defence [matters], our activities must be directed towards the evacuation [of the Arabs] from the whole Beit Shean Valley apart from the town [that is, Beisan]. Now is the time. We shall [later] pay the owners the price of the land.'[16] At Daliyat ar Ruha and Qira wa Qamun, Weitz had dealt with the eviction of Arab tenant-farmers squatting on Jewish-owned lands; in the Beit Shean Valley, he was dealing with the expulsion of Arabs from Arab-owned lands (as well as from Jewish lands) to Transjordan.

At the 26 March meeting with Goldenberg, Weitz also raised the problem of the villages of Tira and Qumiya, near Kibbutz Ein Harod, in the Jezreel Valley. 'These villages are also planted in our midst and their inhabitants do not accept the responsibility of preventing the infiltration' of Arab irregulars. 'They must be forced to leave their villages until peace comes. If they do not agree [to go] they must be forced to do it.'[17] Two days later, Weitz reported that David Baum, the Haganah area commander in Kfar Yehezkeel, near Ein Harod, had told him that the bulk of Qumiya's inhabitants had departed, the British authorities having provided transport. The villagers had left behind 'less than 20 men to look after the village and its fields. However, this cannot be allowed', commented Weitz and 'it is clear that our people will have to take over [the village]'. As to Tira, Weitz wrote, it was still 'an obstacle. There is

[15] CZA, A-246/13, p. 2,315, entry for 20 Feb. 1948.
[16] Weitz p. 257, entry for 26 Mar. 1948.
[17] Ibid., p. 257, entry for 26 Mar. 1948.

still no decision to transfer its inhabitants, but we shall have to do
this. We must . . . We must plug ⸴he holes in our midst, even against
our will', he wrote.[18] Tira was evacuated by its inhabitants on 15
April after receiving 'friendly Jewish advice'.[19] Weitz presented the
Qumiya incident in a somewhat different light to his boss, JNF
chairman Dr Avraham Granovsky (later Granott). On 31 March
Weitz wrote him that 'there is a tendency among our Arab neighbours
. . . to abandon their villages. Last Saturday all the inhabitants of
Qumiya . . . left and went to [the town of] Beit Shean [Beisan]. This
will force our people to enter the village. More such [events] are
imminent. And this poses for us very interesting and very grave
problems'.[20]

By mid-April the Beit Shean Valley problem had not been solved.
Weitz raised the matter on 13 April with the chief of the Haganah
National Staff, Yisrael Galili, citing pressure by Goldenberg. According to Weitz, Goldenberg was demanding Haganah National Staff
permission to take over the village of Al Ashrafiya. Galili, according
to Weitz, promised to reply the following day, but failed to do so.[21]
The Haganah did not occupy the village and on 26 April Goldenberg,
meeting Weitz in Haifa, reported that the Arabs were busy reaping
the crops in the Ashrafiya fields 'under British-army guard'. No
word had reached him from Galili and 'the beduins are beginning to
return. King Abdullah [of Transjordan] is pressing them to go back
[to the Beit Shean Valley]'.[22] Weitz responded that 'we must
employ counter-pressure so that also those who have remained in
the valley will leave it'.[23]

The situation was apparently no better on 4 May, when a delegation
of Beit Shean and Jezreel Valley settlement leaders came to see
Weitz in Tel Aviv. They said that units of the Transjordanian army,
the Arab Legion (then formally under British control) had returned
to the town of Beisan. They asked 'me to influence the [Haganah]
staff here [to launch an attack]'. Weitz for his part, was angry with
the settlement leaders that 'the [Beit Shean] Valley was still seething
with enemies'. He told them: 'The eviction [of the Arabs from]

[18] Weitz p. 257, entry for 28 Mar. 1948.   [19] Private information.
[20] CZA, Eliezer Granovsky Papers, A-202/217, Y. Weitz to E. Granovsky,
31 Mar. 1948.
[21] CZA, A-246/13, p. 2,353, entry for 13 Apr.
[22] CZA, A-246/13, p. 2,366, entry for 26 Apr. 1948.
[23] Weitz, iii, p. 273, entry for 26 Apr. 1948.

the valley is the order of the day . . .' That evening, Weitz went to Ben-Gurion's aide, Shkolnik, and spoke about Beit Shean. 'He agreed with me', Weitz recorded. Shkolnik told him to go and talk to Ben-Gurion and the Haganah brass. Whether Weitz subsequently pressed the matter with Ben-Gurion and the Haganah staff in Tel Aviv is unclear.[24] But during the following weeks, 'counter pressure' was effectively applied and the valley was completely cleared of its Arab inhabitants; the town of Beit Shean fell to Haganah attack on 12 May and most of its inhabitants fled or were expelled across the Jordan River to Transjordan.[25] The remaining inhabitants were expelled a few weeks later. On 10 June, two kibbutzim—later named Reshafim and Sheluhot—were established on Ashrafiya lands. On 25 May, Weitz reported to Granovsky that 'we control all the Beit Shean Valley [and] that most of the Arab population has left, and the town [of Beisan] is almost completely empty: Only some 300 souls are left in it for now'.[26] On 13 June Weitz recorded that no Arabs were left in the town.[27]

The extent of Weitz's influence on events in the Beit Shean Valley in the spring and summer of 1948 is unclear, though it does not seem to have been negligible. His intervention in developments in western Galilee seems to have been even more effective. During his visit to Haifa in late March, Weitz instructed the JNF office officials, regarding the Ghawarina or Al Awarna beduins of the Haifa Bay area: 'They must be evicted from there so that they too will not join our enemies.'[28] The officials apparently discussed the matter with the Haifa area Haganah commanders, but by mid-April nothing had happened.

On 14 April a representative of Kibbutz Kfar Masaryk in the bay came to see Weitz in Tel Aviv and raised the question of 'the eviction of the beduins from the bay. [He] was astonished', recorded Weitz, 'that [it] had not been done yet. I [then] wrote a letter to the Haganah [commander] there and to [Mordechai] Shachevitz [a Haganah intelligence officer and JNF land-purchasing agent in

[24] CZA, A-246/13, p. 2,373, entry for 4 May 1948.
[25] *Ilan Va-shelah*, ed. Binyamin Etzioni, IDF publications Maarachot (Tel Aviv, ?1950), 146.
[26] CZA, S-53/437, Y. Weitz to E. Granovsky, 25 May 1948.
[27] Weitz, iii, pp. 310–12, entry for 13 June 1948.
[28] Ibid., p. 257, entry for 27 Mar. 1948.

Haifa] to move quickly in this matter.'[29] A week later, Shachevitz informed Weitz that 'most of the beduins in the bay [area] had gone, [but] some 15–20 men had stayed behind to guard [the abandoned property]. I [Weitz] demanded that these also be evicted and that the [beduin] fields be ploughed over so that no trace of them remains.'[30]

On 26 April Weitz recorded that the northern part of the bay was completely clear of beduins, their shacks destroyed, and their fields ploughed over. But in the southern end, 'the operation must still be completed . . . we must be rid of the parasites'. But there were pangs of conscience, for Weitz added: 'In war—[act] as befits war.'[31] Meeting the following day with representatives of the two bay kibbutzim, Kfar Masaryk and Ein Hamifratz, Weitz 'ordered them to finish the job within five days'.[32] On 25 May Weitz was able to inform Granovsky that 'our Haifa Bay has been completely evacuated [by the Arabs] and there is almost no trace of those who had trespassed on our lands'.[33]

SETTING UP THE FIRST TRANSFER COMMITTEE

Through his contacts in the Haganah, the JNF offices, and the settlements around Palestine, Weitz through the spring of 1948 helped facilitate the exodus of Arab communities from various localities. But the problem—'the Arab problem'—was a national one, and required a comprehensive solution on a national level. During the spring Weitz began lobbying among the political leadership of the Yishuv for the appointment of an authority or committee which would co-ordinate and oversee what he called 'the Transfer Policy'.

Through March Weitz pressed for an appointment with Haganah national chief Galili 'to discuss the problem of new Jewish settlements and the question of the Arab villages'. But Gailili kept calling off scheduled meetings.[34] The two at last met on 31 March, with Weitz,

[29] CZA, A-246/13, p. 2,354, entry for 14 Apr. 1948.
[30] CZA, A-246/13, p. 2,364, entry for 23 Apr. 1948.
[31] CZA, A-246/13, p. 2,367, entry for 26 Apr. 1948.
[32] CZA, A-246/13, p. 2,368, entry for 27 Apr. 1948.
[33] CZA, S-53/437, Y. Weitz to E. Granovsky, 25 May 1948.
[34] CZA, A-246/13, p. 2,338, entry for 18 Mar. 1948.

as he put it, 'raising the question of the evacuation / eviction[35] of the Arabs from our [that is, November 1947 Partition Plan Jewish] areas, which has begun of its own accord in certain places'. Weitz 'demanded' that a policy be decided upon and 'the appointment of a committee with authority to act'. Galili, recorded Weitz, reacted 'positively' and said that he would bring up the matter with the relevant officials.[36]

That evening Weitz met in Tel Aviv with the members of the newly formed Committee for Abandoned Arab Property, which was composed of senior Arab affairs and intelligence experts, including Gad and Moshe Machnes, two prominent Tel Aviv citrus-grove owners, Ezra Danin, a senior Haganah intelligence officer and leading Hadera businessman, and Yoav Zuckerman, the Gedera-based JNF land-purchasing agent. The question discussed was: should the Yishuv 'help create conditions that will bring about an [Arab] evacuation or the contrary: To influence to avoid evacuation . . .? I myself adopted the first line. Ezra [Danin] agreed with my approach . . . The others hadn't a clear stand. They maintained that they had not been appointed to decide on a policy of eviction but to look after [abandoned Arab] property. Let those empowered to do so deal with my proposal', was the general attitude, Weitz recorded.[37]

Weitz, however, felt the matter too important to let drop; the opportunity must not be allowed to slip away. On 2 April he went to see Shkolnik, who was soon to be named Deputy Minister of Defence. They discussed 'the question of the eviction / evacuation of the Arabs from various [Jewish] areas. He knows nothing at all about it.' Weitz concluded that he would have to deal personally with whatever pertained to JNF lands.[38] Weitz pressed Shkolnik to arrange a meeting for him with Ben-Gurion to discuss 'the problem of the eviction / evacuation of the Arabs'.[39] Ben-Gurion, directing the Yishuv's life-and-death struggle, was apparently too busy.

Weitz persisted in his efforts to obtain a national-level decision. On the evening of 14 April, after spending part of the day dealing with the Haifa Bay beduin problem, Weitz again met with members

[35] The Hebrew word *pinui* means both voluntary evacuation and eviction or expulsion. It is used by Weitz repeatedly. I have translated it in accordance with the context in which it appears, though occasionally there is ambivalence.
[36] Weitz, iii, p. 260, entry for 31 Mar. 1948.
[37] Ibid.   [38] Ibid., p. 261, entry for 2 Apr. 1948.
[39] Ibid., pp. 261–2, entry for 4 Apr. 1948.

of the Committee for Abandoned Arab Property—Danin, Gad
Machnes, and Yitzhak Gvirtz, a member of Kibbutz Shefayim and
a Haganah intelligence officer. According to Weitz's account, the
committee members wanted to know his attitude to their work.
Weitz said he was 'dissatisfied'; they dealt with the safeguarding of
Arab property, not 'political military' decision-making. What was
needed, said Weitz, was a body which 'would direct the Yishuv's
war with the aim of evicting as many Arabs as possible from our
[that is, November 1947 Partition Plan] areas. The preservation of
their property after the eviction is a secondary question in which
one could act with great fairness.' The meeting ended, Weitz
recorded, with the committee members asking him to prepare 'an
eviction proposal'.[40]

Weitz quickly got to work. Four days later he presented his
proposal to the committee. 'I drew up a list of the Arab villages
which I think must be cleared [of Arabs] in order to round out
Jewish areas. I also made a summary of the places which . . . need
to be settled [by Jews] . . .'.[41] Weitz set out his thinking on the
matter in his diary entry for 22 April, when referring to the Haganah
conquest that day of Haifa. The Arab population had begun to flee:
'I think that this state of mind should be exploited . . . and [we
must] hound the rest of the inhabitants so that they should not
surrender [and then stay put]. We must set up our state.'[42]

The next day Weitz went up to Haifa, and on 24 April the
adjutant of the city's Haganah commander came to see him. The
officer told Weitz that that day, two Arab suburbs of Haifa, Balad
esh Sheikh and Yajur, were 'being evacuated', adding that perhaps
'in the coming days the road to the north beyond Acre would be
opened and the [Arab] villages [along the way] will have been
shaken [by Haganah actions]. I was happy to hear from him that this
policy was being implemented by the [Haganah] headquarters [in
Haifa]. To frighten the Arabs [into flight] so long as the flight-
inducing fear was upon them.'[43] The Haganah victories and the
mass Arab exodus of April–May reinforced Weitz's conviction that

[40] CZA, A-246/13, p. 2,354, entry for 14 Apr. 1948.
[41] CZA, A-246/13, p. 2,358, entry for 18 Apr. 1948. In the published version of
the diaries, Weitz altered the phrase 'villages which I think must be cleared [of
Arabs]', to 'Arab villages which have been evacuated', See Weitz, iii, p. 268.
[42] CZA, A-246/13, p. 2,364, entry for 22 Apr. 1948.
[43] CZA, A-246/13, p. 2,365, entry for 24 Apr. 1948.

a fundamental change was taking place in Palestine, 'a complete territorial revolution'.[44] But the transformation had to be controlled, speeded up, and augmented. New life was breathed into Weitz's efforts to obtain a national policy decision from the Yishuv's leaders with Ezra Danin's resignation in mid-May from the Committee for Abandoned Arab Property, which had proved almost completely ineffective in stemming the looting of Arab properties by Haganah, IZL, and Lehi units, and by individuals, in the wake of each conquest.

On 18 May, Danin, home and unemployed in Hadera, wrote Weitz that he was now free and willing to join the type of body Weitz was proposing should be established. Danin explained that he had left the property committee because 'it had not been my purpose to cover up individual acts of robbery'. In any case, he agreed, this was all a relatively irrelevant sideshow. What was now needed was 'an institution whose role will be . . . to seek ways to carry out the transfer of the Arab population at this opportunity when it has left its normal place of residence'. Danin thought that 'Christian interests' could perhaps be found who, acting under the banner of 'saving the refugees', could be persuaded to help resettle them permanently in Arab countries. 'Let us not waste the fact that a large Arab population has moved from its home, and achieving such a thing again would be very difficult in normal times', Danin wrote. He offered some concrete proposals, including buying up Arab land, for which 'the time is ripe'. He wrote: 'If we do not seek to encourage the return of the Arabs to the villages or towns they inhabited—then they must be confronted by *faits accomplis*.' Danin outlined the types of *faits accomplis* he had in mind: 'One possibility is the destruction of the buildings (this needs a central decision, and you promised to deal with this, and nothing substantial has been done in this area). Another [means] is settling Jews in all the area evacuated and to find legal means which will give the new [Jewish] inhabitants certain rights [over the abandoned property] . . .'.[45]

Danin's joining Weitz seems to have added important momentum to Weitz's efforts to persuade the Yishuv leadership to pursue an organized 'Transfer Policy' *vis-à-vis* the Palestinian Arabs. That

[44] Weitz, iii, p. 288, entry for 20 May 1948.

[45] Institute for Study of Settlement (Rehovot), Weitz Papers, E. Danin (Hadera) to Y. Weitz (Tel Aviv), 18 May 1948.

policy was the focus of the meeting between Weitz and the newly appointed Foreign Minister of Israel, Moshe Shertok (later Sharrett) on 28 May. Should the Yishuv 'take action to transform the flight of the Arabs from the country and the prohibition of their return into an accomplished fact? If so, is it not necessary to empower a person or two to three people to deal with this according to a preconceived plan?' Weitz asked Shertok. Weitz then proposed that the Cabinet appoint himself, Elias Sasson, the newly appointed head of the Foreign Ministry's Middle East Affairs Department, and Danin 'to hammer out a plan of action designed [to achieve] the goal of transfer'.

According to Weitz, Shertok 'congratulated' him on his initiative. Shertok's 'view also is that this momentum [of Arab flight] must be exploited and turned into an accomplished fact,' but the Foreign Minister wanted first to consult with Ben-Gurion and Finance Minister Eliezer Kaplan. For the moment, Shertok cautiously endorsed Weitz's proposal that the Yishuv buy land from 'departing Arabs'.[46] Shertok's own notes from the 28 May conversation, while somewhat cryptic, conform with the drift of Weitz's diary entry. Shertok headed his note: 'To discuss with B[en-]G[urion]'. Referring to 'transfer', the Foreign Minister wrote that 'after the fact of evacuation / eviction, exchange of population [Jews from Arab lands would come to Israel as Palestinians departed for Arab lands], appointment of a committee'. Shertok mentions Weitz, Danin, and Sasson as the first three candidates for the proposed committee. According to Shertok, Weitz also proposed the immediate establishment of Jewish settlements in areas evacuated by the Arabs—specific mention is made of western Galilee and the Jerusalem corridor—as well as purchase of land from departing Arabs. Shertok does not mention destroying Arab villages or exacerbating the ongoing Arab exodus as among the measures discussed.[47]

THE FIRST 'SELF-APPOINTED' TRANSFER COMMITTEE

Two days later, Weitz, Danin, and Sasson met to outline the committee's prospective work. 'From now on, I shall call it the

[46] CZA, A-246/13, p. 2,403, entry for 28 May 1948.
[47] ISA, FM 2564/20, memorandum entitled '[Talk] with Yosef Weitz' (in pencil, by hand), 28 May 1948.

"Transfer Committee"', Weitz records in his diary. The official Cabinet appointment of the committee was imminent, he felt. That evening, 30 May, Weitz discussed the matter with Finance Minister Kaplan, who, according to Weitz, agreed 'that the fact of the transfer must be consolidated and that the Arabs not be allowed to return'.[48] On 1 June, a number of senior ministers and officials, including Foreign Minister Shertok, Minority Affairs Minister Shitrit and Cabinet Secretary Ze'ev Sharef, met in Tel Aviv and, according to Ben-Gurion's diary, decided that 'the Arabs were not to be helped to return', and that IDF commanders 'were to be issued orders in the matter'.[49] Weitz, who apparently had not heard from Shertok for about a week, felt ignored and put out. On 3 June he wrote a strong letter of complaint to the Foreign Minister: '. . . You promised me a speedy answer on the question of "retroactive transfer".' Weitz related that he had met with Danin and Sasson, with Shitrit and Kaplan, and had 'prepared an outline scheme. But from this point on we cannot take a single step forward without getting your consent.'[50] The next day, 'the committee that had appointed itself', as Weitz referred to it, met in Tel Aviv to discuss 'the miracle' of the Arab exodus 'and how to make it permanent'. The committee concluded that 'first of all, the return of the Arabs must be prevented'. Money was needed, 'and I agreed to allocate I£5,000 to Ezra [Danin] in order to begin destruction and renovation activities in the Beit Shean Valley, near Ein Hashofet [Ramot-Menashe] and in the Sharon [the Coastal Plain]'. Destruction of the abandoned villages meant that the refugees would have nowhere to return to; renovation meant readying the sites for Jewish settlement.[51]

The next day, 5 June, Weitz at last saw Ben-Gurion. Weitz came armed with a comprehensive, three-page memorandum, entitled 'Retroactive Transfer, A Scheme for the Solution of the Arab

[48] Weitz, iii, p. 294, entry for 30 May 1948.
[49] DBG, ii, p. 477, entry for 1 June 1948.
[50] ISA, FM 2564/20, Y. Weitz to M. Shertok, 3 June 1948.
[51] CZA, A-246/13, p. 2,410, entry for 4 June 1948. In the published version of the diaries, Weitz wrote not that he had allocated I£5,000 (about $20,000 at the time, a not insignificant sum) from JNF funds but that 'we have obtained the sum of I£5,000 . . . .'. Presumably, Weitz in 1965 was loath to reveal that in early June 1948, without JNF directorate authorization, he had allocated JNF funds for the destruction of villages. Subsequently, on 10 June, the directorate voted I£10,000 for Weitz to use 'to cover eviction expenses'. See Weitz, iii, p. 301, entry for 10 June 1948.

Question in the State of Israel', and signed by Weitz, Danin, and Sasson. In the memorandum, the committee stated that the war had brought about an 'unexpected phenomenon': 'the uprooting of masses [of Arabs] from their towns and villages and their flight out of the area of Israel to neighbouring countries . . .'. By 2 June, the committee estimated, some 190 villages and seven towns had been evacuated, encompassing some 335,000 Arabs. 'This process may continue as the war continues and our army advances', the memorandum went on. The flight and its consequences—the destruction and takeover of Arab property—would no doubt deepen Arab enmity for Israel, the committee warned, creating an enmity so strong 'as perhaps to make impossible the existence of hundreds of thousands of Arabs in the State of Israel and the existence of the state with hundreds of thousands of inhabitants who bear that hatred'. Left in place, such a minority would cause Israel no end of problems.

Two months later, Danin was to set down clearly his feelings in this matter in a letter to a critic of the government's hard-line policy towards the Arabs:

I believe with complete faith that the good of both peoples lies in complete separation. Here will remain a deposit of bitterness which will encourage endless conflict. Therefore, I would do everything possible to reduce the number of this minority. . . .[52]

According to Weitz, Danin, and Sasson, the conclusions to be drawn from these assumptions were, first, that 'fair neighbourly' relations, long sought by the Yishuv with the Arabs, 'have been undermined, not through our will . . . Israel must be inhabited largely by Jews, so that there will be in it very few non-Jews.' And second, 'the uprooting of the Arabs should be seen as a solution to the Arab question in the State of Israel and, in line with this, it must from now on be directed according to a calculated plan geared towards the goal of "retroactive transfer" . . .'. The committee then outlined its proposals for 'action' to consolidate and amplify the transformation which was taking place:

(1) Preventing the Arabs from returning to their places.
(2) [Extending] help to the Arabs to be absorbed in other places.

---

[52] ISA, FM 2570/11, E. Danin (Tel Aviv) to Yitzhak Avira, a member of Kibbutz Ashdot Ya'akov in the Jordan Valley, 16 Aug. 1948.

The first activity subdivides into a number of aspects:

> (1) Destruction of villages as much as possible during military operations.
> (2) Prevention of any cultivation of land by them, including reaping, collection [of crops], picking [olives] and so on, also during times of ceasefire.
> (3) Settlement of Jews in a number of villages and towns so that no 'vacuum' is created.
> (4) Enacting legislation [geared to barring a return] . . .
> (5) [Making] propaganda aimed at non-return.

The committee also proposed purchasing land from Arabs leaving the country, negotiating with the Arab countries about a planned resettlement of the refugees, and assessing the value of the property left behind by the refugees. To implement the plan, the committee —which still had no official standing—proposed the establishment of a 'non-governmental three-man committee, invested with clear and defined cabinet authority'. The committee would set up a labour battalion to 'destroy and / or renovate Arab villages . . ., co-ordinate Jewish settlement in the empty sites, prepare legislation to legalize the transformation and to negotiate with Arab representatives the resettlement of the refugees'. The committee estimated that a budget of 'I£150,000' would be needed to cover 'destruction and renovation in village and town', 'I£2,250,000' for setting up some ninety new Jewish settlements and 'I£100,000' to bankroll the committee's apparatus at home and abroad. No mention is made of a budget for funding the refugees' resettlement abroad—though presumably the abandoned property to be expropriated by the Yishuv would be used to cover that.[53]

How did Ben-Gurion react? According to Weitz, 'he agreed to the whole line [that is, policy]', but thought there was an order of priority. According to Weitz, Ben-Gurion wanted 'the actions in the country done first'—destruction of villages, settlement on abandoned sites, prevention of Arab cultivation. Planning the refugees' resettlement abroad could be done later. Ben-Gurion favoured the idea of a supervisory committee but 'he did not agree to our temporary committee'. Nevertheless, Weitz then told the

---

[53] ISA, FM 2564/19, Memorandum entitled 'Retroactive Transfer, A Scheme for the Solution of the Arab Question in the State of Israel', undated, signed Y. Weitz, E. Sasson, E. Danin.

Prime Minister that he had 'already given orders to begin here and there destroying villages and [Ben-Gurion] approved this. I left it at that', Weitz records.[54] According to Ben-Gurion, who devoted two paragraphs in his diary to the meeting, he proposed that a committee of three—composed of representatives of the JNF (Weitz), the Jewish Agency settlement department, and the Agency's treasury department—be set up, its job being to oversee 'the cleaning up of the [Arab] settlements, cultivation of their [fields] and their settlement [by Jews], and the creation of a labour battalion to carry out this work'. Ben-Gurion, like Weitz, stressed that it would not be the government carrying out these activities, but they would be carried out 'with its knowledge, by the National Institutions'. Nowhere in these paragraphs does Ben-Gurion refer clearly to the destruction of Arab villages, the prevention of a return of the refugees, amplifying the Arab exodus, or buying up Arab land as having been broached during the discussion; nor does he refer anywhere, even obliquely, to the Weitz committee memorandum.[55]

The next day, 6 June, Weitz sent Ben-Gurion a detailed list of the abandoned villages and towns, with the appropriate population figures, and a covering note stating, 'I . . . allow myself to set down your answer [yesterday 5 June] to the scheme-proposal I submitted to you, that: (A) You will call a meeting immediately to discuss [the scheme] and to appoint a committee and [outline] its plan of work; (B) You agree that the actions marked in clauses I, 2 . . . will begin immediately [that is, the destruction of Arab villages and prevention of cultivation of Arab fields].' Weitz continued: 'In line with this, I have given an order to begin [these operations] in different parts of the Galilee, in the Beit Shean Valley, in the Hills of Ephraim and in Samaria [meaning the Hefer Valley].'[56] Weitz was covering himself. Aware of Ben-Gurion's *modus operandi* in this sensitive matter— nothing in writing, nothing traceable directly to himself or the Cabinet—Weitz was trying not to leave himself open to charges that he had acted on his own. Perhaps he also hoped that a written response would be forthcoming from the Prime Minister, as is implied by his diary entry for 7 June: 'Have heard nothing from

[54] CZA, A-246/13, p. 2,411, entry for 5 June 1948.
[55] DBG, ii, p. 487, entry for 5 June 1948.
[56] ISA, FM 2564/19, Y. Weitz to D. Ben-Gurion, 6 June 1948. (Also dated 'Year One to the Freedom of Israel.')

D[avid] B[en] G[urion] and M[oshe] S[hertok]. I am acting in this matter, therefore, on my own. Is it right?'

But, at this stage, Weitz was not to be deterred by the lack of a formal, written permit for his activities. He spent the day talking with Danin about how to go about destroying the abandoned villages— where would the money come from, the tractors, the dynamite, the manpower? And where was it best to begin? 'Preparations are under way for action in the villages. We have brought in [Yoav] Zuckerman, who will act in his area [that is, around Gedera, southeast of Tel Aviv]. The questions are many: The town of Beit Shean, to leave it alone completely, or part of it . . . and Acre and Jaffa? and Qaqun? And what shall we work with? We are looking for tools of destruction.'[57]

With most able-bodied men in the Yishuv conscripted into the IDF, with most equipment, such as tractors and tracked caterpillars, in use by the army or in agriculture, and with dynamite in perennially short supply, Weitz had a job of it organizing what amounted to an enormous project of destruction. Added to this was Weitz's unease, from the start, because of the absence of formal, explicit endorsement of his operations by Ben-Gurion, the Prime Minister's oral declarations notwithstanding. Ben-Gurion wanted Weitz to implement the 'Transfer Committee' scheme, but without his name being involved or cited as the source of the committee's authority. It were best that 'things' happened as if of their own—villages razed, cultivators barred from their fields, refugees barred from returning and, indeed, villagers expelled—without his personal imprimatur. Ben-Gurion in this was probably as sensitive to the reputation of the new Jewish state as to his own standing in future histories.

But there is no doubt that Ben-Gurion agreed to Weitz's scheme. Finance Minister Eliezer Kaplan said as much to Weitz when they met on 8 June, adding his own endorsement of the plan. Kaplan added that he had concluded with Ben-Gurion that a committee to oversee the plan's implementation be set up, composed of representatives of the JNF, the Jewish Agency, and the Ministry of Defence. (The inclusion of a Defence Ministry man was an innovation and ran counter to both Ben-Gurion's and Weitz's initial desire that the government not be directly involved, and tarnished in so doing. But

[57] CZA, A-246/13, entry for 7 June 1948.

technically the activities of the committee were better carried out in co-ordination with the IDF—clearly requiring a Defence Ministry official to liaise.) Weitz proposed to Kaplan that Danin, who belonged to none of these bodies, also be included, and Kaplan agreed, Weitz recorded.[58]

Weitz seemed to have made some progress in early June in establishing 'his' committee. But an omen appeared a few days later, when the 'natural' Defence Ministry candidate for the committee, Levi Shkolnik, on 10 June informed Weitz that he was not joining. Weitz sadly recorded this as a symptom of the 'complete chaos which reigns in thinking, or lack of thinking, about this grave subject, as about other subjects', in the political leadership of the Yishuv.[59]

Meanwhile, Weitz's temporary Transfer Committee continued functioning. Weitz sent two settlement officials, Asher Bobritzky and Moshe Berger, to tour the coastal plain 'to determine in which villages we will be able to settle our people, and which should be destroyed'.[60] At the same time, Zuckerman informed him that he had 'arranged the destruction [of the village] of Al Mughar [near Gedera], which will begin tomorrow morning'. That same day the JNF directorate allocated I£10,000 to Weitz to carry out the work of destruction.[61]

On 13 June, Weitz travelled north to the Beit Shean and Jezreel valleys, where he saw 'our people . . . reaping in the fields of [the Arab village of] Zir'in'. In Kibbutz Beit Hashita, Weitz met Goldenberg, David Baum from Kfar Yehezkeel, and the commander of the IDF's (Golani Brigade) battalion in the Jezreel, Avraham Yoffe. 'From the start of our talk', Weitz recorded, 'it became clear that there is agreement among us on the question of the abandoned villages: Destruction, renovation and settlement [by Jews].' Another participant in the meeting, representing Kibbutz Nir David, urged the immediate settlement of 'Zir'in, Sandala and Faqqu'a' and barring a return of the refugees to the Beit Shean Valley 'so long as no peace treaty was signed'. Weitz concluded: 'There is a consensus

    [58] Weitz, iii, p. 300, entry for 8 June 1948.
    [59] CZA, A-246/13, p. 2,415, entry for 10 June 1948.
    [60] Weitz, iii, p. 301, entry for 10 June 1948 and CZA, A-246/13, p. 2,415, entry for 10 June 1948.
    [61] Weitz, iii, p. 301, entry for 10 June 1948.

everywhere: "No" to a return of Arabs, on no account. They must be prevented from returning and at the same time the "vacuum" must be filled [with new Jewish settlements].'[62]

On his way back to Tel Aviv the following day, Weitz met Danin, who reported on a conversation he had had (apparently on 13 June) with Foreign Minister Shertok about the Transfer Committee's work. 'It appears that he [that is, Shertok] has not yet taken a [formal, open] position', wrote Weitz. Shertok reportedly asked Danin: 'Are you doing [things]?' Danin replied: 'Yes'. Shertok said: 'Continue to do [things].' Danin then asked the Foreign Minister what he would say when asked abroad about the committee's activities. Shertok 'remained silent', Danin reportedly told Weitz.[63] During their meeting, Danin also told Weitz about the progress made in the destruction of the village of Fajja, near Petah Tikva.[64] Zuckerman that day gave Weitz a progress report on the destruction of Al Mughar.[65] The next day, Weitz went to see for himself. At Al Mughar

three tractors are completing its destruction. I was surprised [as] nothing moved in me at the sight . . . Not regret and not hatred, as this is the way of the world. Yes, we [the Jews, now] want to enjoy life in this world and not in the next. We simply want to live, and the dwellers of these mud-houses did not want us to exist here. They sought not only to rule over us but also to destroy us . . . And, interestingly, this is the opinion of all our boys, from one end [of the political spectrum] to the other. If there are doubts about this work, then they are among certain political party leaders who have not forgotten to employ their dialectics . . . [The reference is to the Marxist Hashomer Hatza'ir element in Mapam].[66]

On 16 June, probably on the basis of a progress report from Weitz, Ben-Gurion summarized the destruction of some of the Arab villages to date:

[Al] Mughar near Gedera, Fajja, Biyar Adas (near Magdiel) have been destroyed. [Destruction is proceeding in] Miska (near Ramat Hakovesh), Beit Dajan (east of Tel Aviv), in [the] Hula [Valley], [in] Khawassa near Haifa, As Sumeiriya near Acre and Ja'tun [? perhaps Khirbet Ja'tun] near

[62] Ibid., entry for 13 June 1948.
[63] Ibid., p. 302, entry for 14 June 1948.
[64] CZA, A-246/13, p. 2,418, entry for 14 June 1948.
[65] Ibid.
[66] Weitz, iii, p. 303, entry for 15 June 1948.

Nahariya, Manshiya . . . near Acre. Daliyat ar Ruha has been destroyed and work is about to begin at [Al] Buteimat and Sabbarin [both in the Ramot-Menashe area].[67]

Meanwhile, Bobritzky, flanked by Moshe Berger and another of Weitz's Haganah Intelligence Service and JNF land-purchasing contacts, Yaakov Barazani, spent mid-June touring the coastal plain to determine which sites were suitable for Jewish settlement and which should be levelled. Bobritzky consulted with, and was assisted on his tours by, local IDF commanders. Berger was apparently named to oversee the destruction in the area (as Zuckerman was responsible for operations in the area south of Tel Aviv).

On 11 June, a senior officer in the IDF unit responsible for the coastal plain (from Tel Aviv northwards to Tantura, south of Haifa), the Alexandroni Brigade, informed IDF General Staff / Operations that 'according to the directive by the Arab Department [*sic*], all the Arab villages in the [area of the] brigade must be destroyed after consulting with the Agricultural Centre and the [Jewish] settlement [bodies]. Motta [Moshe] Berger has been appointed to the post [of overseeing the destruction of these villages]. The destruction of these villages will be carried out after permission [is given] by the settlement bodies and by you [that is, IDF General Staff / Operations], from a security viewpoint.'

What General Staff / Operations thought of this letter, and how it responded to it, is unknown. But it highlighted the mechanics of Weitz's operation during June. While attempting to obtain national-level authorization and sanction for his activities, he proceeded to carry out his 'mission' with the tools and personnel at hand, and this included co-ordination with and a measure of instructions to, the local, middle-level IDF authorities.

Bobritzky, who submitted a progress report on his activities on 21 June to Weitz and the other relevant Jewish authorities (the Agricultural Centre and the Jewish Agency Settlement Department), named both Berger and Barazani as being 'charged with destroying the Arab villages'. Bobritzky listed 20 villages which he thought were of no use for Jewish settlement and could be destroyed— including Qannir, Kafr Qari, Bureika, Khirbet as Sarkas, Wadi Ara, the dwelling places of Arab al Fuqara and of ad Dumeira, near

---

[67] DBG, ii, pp. 523–4, entry for 16 June 1948.

Hadera, and Umm Khalid (which Bobritzky misnames as Umm Walid), near Netanya.

He listed seven village sites as suitable for Jewish settlement—including Fajja, which he found was being destroyed and where he asked that '10–11 rooms' be left intact for Jewish settlers, Tantura, Khirbet Beit Lid, An Nusseirat, Sabbarin, and two smaller sites. Four sites, according to Bobritzky, required 'special clarification'. Another eleven—including Miska, Beit Dajan, Arab Abu Kishk, Al Kheiriya, Jalil, and Khirbet Azzun (Tabsar)—he had not yet visited.

In his concluding remarks, Bobritzky thought it worth mentioning that Berger and Barazani 'had orders' to speed up the destruction of the villages and that the IDF district OCs (*mefakdei nafot* or *mafanim*), concurred with this approach. Bobritzky, for his part, asked the two officials and the district OCs to hold off on the destruction of the sites he regarded as appropriate for Jewish settlement. Here, perhaps for the first time, surfaced the inherent contradiction between the settlement officials' desire to preserve buildings and space for the settlement of Jewish immigrants and the political desire, seen by its proponents as a need, to destroy as many habitable sites as possible to prevent an Arab refugee return.[68]

[68] ISA, MAM 307 gimel/33, Bechor Shitrit to David Ben-Gurion, 23 June 1948. Shitrit's letter quotes verbatim the full text of the letter of 11 June 1948 from 'Oded', Alexandroni Brigade to General Staff / Operations. I would like to thank Dr Charles Kamen for letting me see a copy of this document. For Bobritzky's report, see LA 235 IV/2060 aleph, A. Bobritzky to Settlement Department, 21 June 1948. All three letters, from Minority Affairs Minister Shitrit, from the Alexandroni Brigade officer 'Oded' and from the settlement official Bobritzky, reflect the bureaucratic decision-making confusion of the time as well as the semi-clandestine nature of Weitz's Transfer Committee activities. The State of Israel had only just come into being, and under the worst of circumstances. Large areas remained in which authority was unclear or disputed. The Ministry for Minority Affairs was only donning skin and bone; the Arab Affairs Committee of the National Institutions (of which Weitz, Danin, and Sasson were members) and Gad Machnes's Abandoned Property Committee were becoming powerless, empty hulks. Weitz's Transfer Committee, operating in the shadows, still lacked official sanction and public recognition. The Ministry of Agriculture was only beginning to flex its muscles and usurp some of the authority traditionally vested in the Settlement Department of the Jewish Agency, the Agricultural Centre, and the JNF Lands Department. In general, the line demarcating the IDF's authority from the various civil agencies—in all that concerned the Arab minority and the abandoned Arab villages and property—during the summer of 1948 remained unclear, with all concerned jockeying for positions of authority. It was, all in all, a boom-time for private, semi-official, and official initiatives by single-minded, dogged executives—such as Weitz.

OPPOSING THE TRANSFER COMMITTEE

Ben-Gurion may have been content, indeed happy, with the situation;
the Transfer Committee was operational though without formal
government authorization. But Weitz was growing increasingly
frustrated and perturbed. Months of lobbying had failed to secure
for the Transfer Committee the coveted letter of appointment. On
15 June he again met with Shertok and 'demanded . . . that the
committee be set up at long last with the clear authority to carry out
its complete programme'. According to Weitz, Shertok again
promised to bring up the matter with Ben-Gurion and 'to speed up
the appointment of the committee'.[69] But Ben-Gurion procrastinated.
Since mid-May, the Prime Minister has been busy fending off
embarrassing questions and charges in Cabinet concerning the
destruction of Arab villages. Strong hints were repeatedly thrown
out, especially by Mapam ministers, that this destruction was part of
an unannounced general policy of expulsion against Israel's Arab
population being carried out by Ben-Gurion, his emissaries, and
the IDF behind the Cabinet's back.

During May and the first half of June, Mapam's main political
bodies, the Political Committee and the Centre [*Merkaz Mapam*],
debated policy towards the Palestinian Arabs. On 15 June the
Political Committee summarized the debate with a list of conclusions,
broadcast a week later by the party centre in a circular to all
'party activists'. In forthright language, the statement declared the
party's opposition to the expulsion of Arabs, proposed that the
government issue an appeal to 'peace-loving Arabs' to stay put, and
opposed 'the destruction of Arab villages . . . not out of direct
military necessity'. The statement supported the return of the
refugees to Israel at the end of the war.[70]

Once this position was reached and publicized by Mapam, any
open endorsement by Ben-Gurion or the Cabinet majority of a
policy of 'transfer' would have led to a breakdown of the government
coalition, in which, for all their differences, Ben-Gurion's Mapai
and Mapam were the chief and generally aligned components. The

---

[69] CZA, A-246/13, p. 2,419, entry for 15 June 1948.

[70] HHA-ACP 10.95.11 (1), 'From the Log-Book of the Secretariat, No. 4'. 'Our
Policy Towards the Arab During the War' (The Decisions of the Political Committee
of 15 June 1948), 23 June 1948, Tel Aviv.

appointment by Ben-Gurion or the Cabinet of a committee with terms of reference as proposed by Weitz would have placed Mapam's leaders in an impossible position; the party, whose rank and file in the Kibbutz Artzi and Kibbutz Meuhad settlement movements and urban branches provided much of the IDF's officer corps, most of the country's farmers, and many senior civil servants, would most probably have bolted the coalition.

Ben-Gurion, busy directing a life-and-death struggle against the Arab states, faced with dissension which bordered on revolt from the Right and an open rift with the IDF brass, and under heavy pressure from the UN and the Western powers on a variety of crucial issues, could hardly contemplate a cabinet breakup in mid-1948. Moreover, it ran against Ben-Gurion's grain openly to endorse and put his name to a policy which contemporary and future critics might brand as morally questionable.

It is worth tracing specifically the emergence of awareness among Mapam's leaders of Weitz's essentially covert activities and how this awareness was translated into the criticism and opposition in Cabinet which ultimately frustrated Weitz's hopes of winning official endorsement for the Transfer Committee's activities. At the 26 May meeting of Mapam's Political Committee, Eliezer Prai, a member of Kibbutz Merhavia and the editor of the party's daily newspaper, *Al Hamishmar* (On Guard), charged that there were elements in the Yishuv carrying out a 'transfer policy', by 'blood and fire', aimed at emptying the Jewish state of its Arab inhabitants. 'It has already been said that Weitz gave an order to expel the Arabs from Western Galilee', said Prai, probably referring to Weitz's instructions to the (Mapam-affiliated) kibbutzim Kfar Masaryk and Ein Hamifratz, and possibly also to his 'advice' to the Haganah brass in Haifa, regarding the Arabs in the areas north of Haifa. 'This is the policy and thinking behind [the destruction of the Arab villages in the area]', he said. Another participant at the meeting referred critically, in passing, to Weitz's trips around the country, dispensing 'instructions' regarding the disposition of Arabs and Arab villages.[71]

There was also criticism of a more personal order. In July, Haim Cafri, of Kibbutz Maanit, near Hadera, and a leading Mapam

[71] HHA 66.90 (1), protocol of the meeting of the Mapam Political Committee, 26 May 1948.

defence figure, at a closed meeting of Mapam defence personnel, charged—but without openly naming him—that Ezra Danin had profited personally from the Arab exodus, and that Arab friends of Transfer Committee members were receiving preferential treatment from the Yishuv's executive bodies. Cafri named Danin and (mistakenly) Gad Machnes (who was by then the Director General of the Minority Affairs Ministry) as the men responsible for 'overseeing the glorious department of "landscape improvement", as it is called in their language'.[72] Mapam's Cabinet ministers, Agriculture Minister Aharon Zisling and Labour Minister Mordechai Bentov, participated in these deliberations, and subsequently raised Mapam's concerns at Cabinet meetings. While previously they had somewhat loosely criticized the destruction of Arab villages, at the Cabinet meeting of 16 June Bentov asked the Prime Minister directly whether there existed a committee, composed of 'E. Danin, Y. Weitz and [sic] M. Nahmani . . . empowered to damage Arab property'.

Ben-Gurion parried, promising 'to investigate and respond'. How, if at all, Ben-Gurion answered the question at a later date is unknown; events during the summer of 1948 came thick and fast, one concern quickly replacing others in ministers' minds. The ministers grew accustomed to Ben-Gurion's failure to answer Cabinet questions.[73]

But Zisling was not to be put off this time. Later in the day, he wrote an angry note to Ben-Gurion, with copies to all Cabinet members, denouncing the depredations against Arab property. 'There is increasing information', he hinted darkly if vaguely,

that [these are] activities which according to their nature and scope cannot be done except by [permission of higher] authority. I would like to know: Who gave the order to demolish houses and destroy settlements—in Beit Shean, in the villages around it and in other parts of the country?

Is there a basis for the rumours that on various occasions responsible persons were heard to give 'suggestions' which were understood by the hearers to be tantamount to an order, giving an assessment of the number of Arabs who were to be left in future in settlements being evacuated

---

[72] HHA 10.18, protocol of meeting of Mapam defence activists, 26 July 1948.
[73] KMA-AZP 9.9.1, 'Decisions of the Provisional Government', meeting of 16 June 1948.

(Haifa and elsewhere)? [The reference is to a report then circulating in Mapam leadership circles that Ben-Gurion had told his aides that he wanted no more than 15,000 Arabs left in Haifa]. If there is a foundation for the information reaching me, [continued Zisling,] then the responsibility stems from government sources, and it is inconceivable that . . . actions of this sort continue contrary to what is heard in cabinet deliberations.[74]

Such criticism helped to stop Ben-Gurion from issuing the letter of appointment sought by Weitz. But it failed to prompt the Prime Minister to issue hard and fast instruction that the Transfer Committee halt its work of destruction immediately. On 14 July Zisling returned to the fray, charging that some IDF officers were blowing up villages 'on the basis of unauthorized orders'. The minister explained that he had heard that 'civilians' were dispensing such orders, citing such an order issued to the IDF commander in the town of Beit Shean. Ben-Gurion interjected: 'A [military] commander cannot receive an order from any committee but only from his [military] superior.' As if paying no attention to the interruption, Zisling went on to recount that he himself had seen Arab villages in the Hadera area being destroyed after the fighting in the area had long died down. Those responsible should be disciplined, Zisling said and, turning to Ben-Gurion, added: 'And I don't accept your answer that you don't know who destroyed them.'

Ben-Gurion challenged Zisling to name such destroyed villages. Zisling replied: '[Khirbet as] Sarkas', a village in the Hadera area whose inhabitants had been expelled by the Haganah on 15 April.[75] Zisling added: 'In an earlier [Cabinet] meeting I mentioned [Moshe] Goldenberg, who spoke of [Weitz's] order to destroy villages and mentioned your [that is, Ben-Gurion's] name, and I said that I did not believe that [Weitz's order] was given in your name.' The implication of Zisling's statement was that whereas in the past he had not believed that Weitz had been acting with Ben-Gurion's authorization, he now suspected that this was the case. Still, he

---

[74] ISA, FM 2401/21, A. Zisling to D. Ben-Gurion, 16 June 1948.

[75] KMA-AZP 9.9.3, transcript of Zisling's statements in the Cabinet meeting of 14 July 1948, being verbatim extracts of the Cabinet minutes. Zisling's reference appears to have been to the meeting on 13 June at Kibbutz Beit Hashita, Jezreel Valley, between Weitz and regional IDF commanders, including Goldenberg, the Beit Alpha area commander, on policy towards the Arabs.

held back from saying so openly, because, possibly, to do so—to
recognize openly that transfer policy stemmed from the Prime
Minister—would leave the Mapam ministers little choice but
to declare an open rift with Mapai and perhaps bolt from the
government.

Weitz and his circle clearly recognized Mapam as the chief
opponents in the Yishuv of a policy of transfer. But Weitz does
not seem to have been aware of the major influence the party and
its representatives in the Cabinet succeeded in exerting on Ben-
Gurion, repeatedly frustrating official endorsement of the Transfer
Committee. Weitz seems not to have known that his committee's
activities were a subject of open Cabinet debate. Mapam's criticisms
of what they regarded as the transfer policy were usually delivered
in closed, or very limited forums, and were never publicly enunciated,
for fear of handing Israel's Arab enemies political ammunition.
Weitz heard Mapam's views chiefly from Kibbutz Artzi movement
leader Yaakov Hazan, who sat with Weitz on the JNF board of
directors. On 9 June, for example, Hazan told Weitz that he
opposed 'transferring all the Arabs to neighbouring countries'.
But, according to Weitz, Hazan agreed to JNF allocation of
funds 'for expenses for destroying villages . . . located on [JNF-
owned] land or on land owned by effendis [that is, usually
absentee, large landowners] which would in the end be sold to
the JNF'.[76]

On 17 June Weitz angrily noted a speech by Mapam leader Meir
Yaari to the Zionist Executive, in which Yaari condemned 'the
eviction of Arab villagers'. Weitz commented in his diary: '. . . as if
he does not know that all his friends in the kibbutzim are doing it
with complete devotion. That is the power of an abstract ideology
[that is, Marxism].'[77]

Weitz's opponents were not restricted to Mapam. His boss in the
JNF, its chairman Avraham Granovsky, a founder of the centrist
Progressive Party, repeatedly attacked at meetings of the JNF
directorate and in private conversations with Weitz, the general
policy of destroying Arab villages as 'a negative and dangerous

[76] Weitz, iii, p. 300, entry for 9 June 1948.
[77] CZA, A-246/13, p. 2,420, entry for 17 June 1948 and Weitz, iii, p. 309, entry for
28 June 1948.

phenomenon'.[78] But Granovsky' opposition seems to have had only a limited effect on Weitz's freedom of action and on his use of JNF funds and facilities for Transfer Committee operations; the majority of members of the JNF board of directors supported Weitz's Mapai-backed policy.

Much more dangerous to his activities was the opposition of Minority Affairs Minister Bechor Shitrit, who saw himself as an oriental-type patron and benefactor of those whose affairs and interests he was charged with overseeing. Weitz met him on 16 June, apparently after the day's Cabinet meeting, and recorded: 'There is no stability in his opinions. They [Shitrit and his ministry director general, Gad Machnes] support me to my face, but I am not sure they don't speak differently behind my back.'

On 23 June[79] Shitrit complained directly to the Prime Minister about the activities of the Weitz committee (though it is unclear whether, at this time, he was fully aware of their extent and of Ben-Gurion's complicity in them). Enclosing a copy of the letter from 'Oded', Alexandroni Brigade, to IDF General Staff / Operations of 11 June (see note 68), Shitrit complained that the instruction to destroy the Coastal Plain villages had not come from 'the Arab Department', which he took to refer to the Minority Affairs Ministry. 'The Arab Department never issued this order or a similar order or even [issued] a hint to destroy Arab villages.' Shitrit bristled at the assertion that the Agricultural Centre and the settlements should be consulted about such destruction. And he demanded that Berger, charged with overseeing the destruction, according to the Alexandroni Brigade letter, be ordered 'not to interfere in matters pertaining to the Arab Department [that is, the Ministry for Minority Affairs]'. Shitrit demanded that Ben-Gurion find out on what basis Alexandroni's 'Oded' had written the letter. Lastly, he asked that

[78] Weitz, iii, p. 310, entry for 1 July 1948. Weitz apparently had problems, albeit ephemeral, with his own staff. Yosef Nahmani, the JNF land office head in the Galilee, on 6 May complained to Weitz about 'the brutal behaviour of our people towards all the Arabs'. Weitz, after persuading Nahmani of the justice of the Zionist cause and the lack of any alternative course of action, jotted down in his diary: 'He apparently finds it difficult to cut himself off from the concepts that we became accustomed to during the period of enslavement under the British Mandate.' CZA, A-246/13, p. 2,377, entry for 6 May 1948.

[79] CZA, A-246/13, pp. 2,419–20, entry for 16 June 1948, and ISA, MAM 307 gimel/33, B. Shitrit to D. Ben-Gurion, 23 June 1948.

everyone else be barred from all activity relating to his ministry's areas of jurisdiction.

By the end of June, the momentum of the first, 'self-appointed Transfer Committee' had fizzled out. 'There are no tools and no materials' with which to continue the destruction of the villages, Weitz recorded.[80] Lack of official recognition entailed lack of funds.

But it went deeper than that. How could Weitz and his committee take upon themselves such politically momentous actions without clear-cut endorsement from the Yishuv's political leaders? Weitz got cold feet. He understood Ben-Gurion's dilemma, and the Prime Minister's tactics, of having others (Weitz, Jewish settlements, local IDF units) carry out the razing of the villages without himself being directly involved in these activities. Weitz saw this as indicating lack of determination and strength on the Prime Minister's part. On 30 June Weitz briefly met Ben-Gurion, who asked about 'the state of the [Transfer Committee] activities'. But the Prime Minister failed to keep his appointment for the following day with Weitz, when Weitz had hoped to settle Transfer Committee matters with the Prime Minister once and for all. Angry and frustrated, Weitz 'instructed [his officials] to cease work'.[81]

Under pressure from the soft-liners, Ben-Gurion had thus brought about a halt to the Transfer Committee's operations, effectively ending the first chapter in its history. Subsequently he informed Foreign Minister Shertok, the committee's main patron in the Cabinet, of the new situation. But neither Prime Minister nor Foreign Minister regarded the situation as final. In mid-July, Shertok issued a 'letter of appointment' to Ezra Danin, naming the Arabist a special adviser to the ministry's Middle East Affairs Department and the ministry's representative 'to the body which will deal with the problem of the transfer of Arabs out of the area of Israel and their resettlement in the neighbouring countries. This body has not yet been set up because the policy of transfer has not yet received the required endorsement. From my conversation with the Prime Minister on this matter', the letter of appointment continued, 'it became clear that the previous directive [that is, instructing Weitz to destroy the villages] has been rescinded and one must now await

---

[80] CZA, A-246/13, p. 2,425, entry for 25 June 1948.
[81] Weitz, iii, p. 310, entry for 1 July 1948.

new directives.'[82] Because of the mounting political pressures, which included a shuttle around the Middle East by UN Mediator for Palestine Count Folke Bernadotte (who was attempting to persuade Israel to allow the refugees back), the IDF's parallel razing of Arab villages was also temporarily curtailed.[83]

But the thrust of the Prime Minister's thinking, Mapam's and Bernadotte's importunings notwithstanding, remained firm and clear. As the second truce in the hostilities approached, Ben-Gurion jotted down his priorities for the weeks ahead. They included strengthening the army, expanding Israel's arms-production capacity, and drawing up 'a list . . . of abandoned villages standing on dangerous, strategic sites, in order to destroy them, and settlement [by Jews] of abandoned places . . . must be taken in hand'.[84] At Mapam's urging, responsibility for the fate of the abandoned villages —destruction, renovation, resettlement—in mid-July devolved upon the newly constituted Ministerial (or Cabinet) Committee for Abandoned Property, which apparently met for the first time on 13 July and was formally empowered by the full Cabinet on 21 July to decide the fate of the villages. The committee was composed of Defence Minister Ben-Gurion, Foreign Minister Shertok, Finance Minister Kaplan, Justice Minister Felix Rosenblueth (later Pinchas Rosen), Minority Affairs Minister Shitrit, and Agriculture Minister Zisling. The *modus operandi* of the ministerial committee was outlined to Weitz on the evening of 21 July by Kaplan: 'The military staff brings its proposals, to destroy or occupy a village, to establish a permanent hold on land, to the committee . . .'. The committee's decision, if unanimous, becomes policy. If there is dissent, the matter is resolved in the full Cabinet.

Authority for destroying the villages had thus been taken out of Weitz's hands. Now the Cabinet meant to abrogate the powers of the various settlement bodies as well. Until mid-July 1948, the Jewish Agency, the JNF, and the Agricultural Centre of the Histadrut, in consultation with the defence forces, had determined settlement policy and had overseen its implementation. 'According to the [new

---

[82] ISA, FM 2,570/1, M. Shertok to E. Danin, 16 July 1948.

[83] KMA-AZP 9.9.1, for the IDF General Staff / Operations order of 6 July 1948 prohibiting 'destruction of towns and villages . . . without special permission or explicit order from the Minister of Defence . . .'.

[84] DBG, ii, p. 603, entry for 21 July 1948.

decision], we can no longer decide on our own . . .', Weitz
complained, 'We may as well pack up', Weitz told Kaplan, blaming
the Mapam ministers, 'the "just men" in our camp, Zisling and
Bentov, who "worry about" the neighbours living in our house'.[85]

BUYING UP ARAB LAND

During July, various kibbutzim continued to turn to Weitz for
advice and help in matters concerning the abandoned villages.
Representatives from Degania Aleph and Kinneret, for example,
came to Tel Aviv to inform him that the regional Jewish settlements
committee had decided against allowing the Arabs to return.
The representatives came to work out 'a settlement plan for the
[abandoned] lands' along the banks of the Jordan River and the Sea
of Galilee. Weitz promised to do what he could.[86] But Weitz's main
attention, diverted from his initial preoccupation with increasing
the Arab exodus and preventing the refugees' return by destroying
their villages, now focused on another aspect of his original transfer
scheme—the purchase of land from the refugees and the financial
inducement of other Arabs to leave their areas and settle outside
Israel.

The day after his angry meeting with Kaplan, Weitz began
planning a trip by Danin and the lawyer Yosef Strumza to France or
Switzerland 'to meet Arabs . . . to buy land (and to investigate the
possibility of [accomplishing] the transfer, I add to myself)'. That
day, Weitz spoke to Zuckerman and Yosef Nahmani, the head of
the JNF land office in the Galilee, about renewing land purchases
from Arabs.[87] On 26 July Weitz wrote a comprehensive survey of
lands the JNF had its eye on for future Jewish settlement and
regional development.[88] The following day, Weitz heard from one
of his purchasing agents about efforts to buy land from the villagers
of Al Hamidiya (on the northern edge of the Beit Shean Valley),
who would then 'move to settle in Transjordan'. 'The question is',
wrote Weitz, 'should we enter into [such a] partial transfer activity
or should this be postponed until the transfer question is solved in its

[85] CZA, A-246/13, p. 2,444, entry for 19 July 1948.
[86] CZA, A-246/13, p. 2,452, entry for 28 July 1948.
[87] CZA, A-246/13, p. 2,445, entry for 22 July 1948.
[88] CZA, A-246/13, p. 2,449, entry for 26 July 1948.

entirety? And if [we should go in for] partial [solutions]—[should we do so] now, when the Arabs who are leaving are still standing on their own two feet, or should we wait until they collapse [and are willing to take lower prices for their land]?'[89] Weitz encountered some unexpected opposition to the Hamidiya land purchase. JNF chairman Granovsky asked: 'Has it already been decided that [the] Beit Shean [Valley] will be clear of Arabs?' But if the purchase was to be made, he added, then certainly the Arabs' prices should be lowered. 'I chuckled to myself', Weitz recorded.[90] Weitz raised the larger issue when the two continued their discussion the following day: the JNF must not wait for instructions from others, even from the Cabinet, in carrying out its historic mission, 'redeeming' the Land of Israel: 'We must act as we have always acted: gain possession of land in all ways, and not formulate new theories which will delay us . . .'. Granovsky apparently agreed.[91]

But purchasing land from Arabs in the circumstances of late summer 1948 was not easy. Nahmani thought that the Arabs in the Galilee and the refugees now in Lebanon would not dare sell their land to Jews; in the circumstances, it was tantamount to high treason. Moreover, the Palestinians in the north in mid-1948 had not yet grasped that the situation had radically altered, and that the refugees would not be returning. The Arabs, complained Weitz, continued to conduct land negotiations along pre-1948 lines, with 'wearying talks, high prices and other conditions'. The Arabs, he thought, must be made to understand that there would be no return and that only by selling to the Jews would they be assured of 'saving their property' and obtaining the means for resettling comfortably elsewhere.[92]

Weitz and his purchasing agents pressed on. After knocking down the Arabs' prices, deals were concluded in September in Hamidiya (140 dunams—a dunam is approximately a quarter of an acre) and Yafia (200 dunams), and in the Hula Valley in Northern Galilee (200 dunams).[93] But Weitz realized that this was chicken-feed. It was 'quite sad', he recorded. 'Among the Arabs there is not even a

[89] CZA, A-246/13, p. 2,451, entry for 27 July 1948.
[90] CZA, A-246/13, p. 2,452, entry for 29 July 1948.
[91] CZA, A-246/13, p. 2,453, entry for 30 July 1948.
[92] Weitz, iii, pp. 323–4, entry for 3 Aug. 1948.
[93] CZA, A-246/13, p. 2,478, entry for 8 Sept. 1948.

sign of panic movement towards selling land. They are apparently
sure that everything will return to its former state . . .'[94]

The matter continued to vex Weitz.

The Arabs do not see their situation as it really is . . . A refugee Arab
dwelling for months now in a sack-tent is not ready to sell his land [near
Nazareth], except for a small amount [of land] in order to maintain his
family and himself. He hopes that soon he will return to cultivate his
holdings as before. He does not consider the other possibility that perhaps
he will not be able to return [to his village], and that it is better for him to get
rid of his property while its value has not completely depreciated.

Scouting around for an explanation for the refugees' stubbornness,
Weitz conjectured: 'Maybe . . . it lies in his rootedness. After all,
he and his kind have been bound to the soil . . . for generations, and
it is not easy to dissolve these ties.'[95]

The publication in September of the 'Bernadotte Plan', the UN
mediator's testament, as it were, to the warring tribes of Palestine
(he was assassinated that month in Jerusalem by Jewish terrorists),
only added to Weitz's difficulties. According to JNF officials in the
north, the plan's provision for the refugees' return promoted a
feeling among the Arabs 'that they were about to return . . . and
[now] felt no compulsion to sell'.[96]

But there were countervailing factors: the political uncertainty,
the continuing expansion of the Jewish state, so clearly stronger
than its Arab enemies, and the Israeli authorities' attitude and
policy towards the Arabs who had remained within the state. At
Yafia, for example, the villagers were willing to sell another 500
dunams but were demanding what Weitz regarded as an excessive
I£25 per dunam. He advised his officials to wait, reasoning that the
Arabs were not being allowed to cultivate their fields; eventually
they would 'be forced to sell'.[97] At the same time, Weitz formulated
plans to buy land from absentee and refugee landowners living in
Cairo, Beirut, and Europe. Foreign Minister Shertok supported
Weitz's scheme to send land-purchasing agents to Paris. But Weitz
wondered out loud whether some of the money paid for land might
not be used by the Arabs to finance their war effort. Shertok

[94] Weitz, iii, p. 339, entry for 9 Sept. 1948.
[95] Ibid., p. 341, Weitz to Rema, 11 Sept. 1948.
[96] CZA, A-246/14, p. 2,497, entry for 29 Sept. 1948.
[97] CZA, A-246/14, pp. 2,537–8, entry for 2 Dec. 1948.

replied: 'The reasons for buying [Arab land] outweigh [the reasons against].'[98]

After much deliberation and delay, Zuckerman and Gad Machnes were eventually shipped off to Paris to seek out Arab landowners and conclude deals. But no sooner were they in Europe than the Cabinet ordered the JNF to stop all land purchases from refugee and local Arabs. Ben-Gurion told Weitz and Danin at a meeting in Tiberias on 18 December: 'The JNF [henceforward] would buy land only from the State. There was no need to buy land from Arabs.' According to his diaries, the Prime Minister argued that Israel's problem was not an absence of land on which to settle Jews but an absence of Jews willing to settle on the land.[99] According to Weitz, Ben-Gurion said that 'the Cabinet was resolved to forbid [the JNF] from buying land from Arabs'.[100] Weitz then ordered the JNF men in the north to suspend purchases 'for the time being'.[101]

Weitz explained the opposition to buying land from Arabs thus: '. . . the politicians . . . say: There is no need to buy land; after all, it was redeemed with the sword and the blood of our sons, and it is [now] all in our hands . . .'.[102] Ben-Gurion preferred to act as did most nations at war—what nation paid for territory after conquering it? Besides, Israel was poor. Weitz cabled Zuckerman and Machnes —who reported that they were on the verge of success—to fly home.[103]

The government's ban on land-purchasing from Arabs threatened the JNF's *raison d'être*. The JNF directors debated the matter on 4 January 1949. The directorate expressed 'understanding' for the Cabinet's opposition to purchases from refugees 'but does not agree to the government's stand' prohibiting also purchase from Arabs living inside Israel. In defiance of the Cabinet, the directorate instructed Weitz to go ahead with purchases of some 55,000 dunams which the JNF had been interested in before the war 'in Shfar-am' [Shafa Amr], Isfiya, Daliyat [al] Karmil, Ijzim, [Al] Birwa, Ilut, Yafia, [Al] Mujeidil, Umm al Kannem [?], Kislut [? Islut], Indur

[98] Weitz, iii, p. 338, entry for 6 Sept. 1948.
[99] DBG, iii, p. 885, entry for 18 Dec. 1948.
[100] Weitz, iii, p. 366, entry for 18 Dec. 1948.
[101] Ibid., and CZA, A-246/14, p. 2,553, entry for 18 Dec. 1948.
[102] Weitz, iii, p. 362, Weitz to Rema, 9 Dec. 1948.
[103] Weitz, iii, p. 373, entry for 28 Dec. 1948, and Weitz iv, p. 3, entry for 6 Jan. 1949.

(Ein Dor), Tamra, [Al] Manara, Beit Shean [Beisan], Kafr Misr,
Sirin, Dabburiya, Maghar, Lubiya, Madhar, Maghar Durruz, Turan,
Arraba, Sakhnin, Deir Hanna and Bueina'.[104]

SETTLING THE ABANDONED LANDS

Weitz resumed his land-purchasing campaign, though it was to
remain his secondary activity during the latter months of 1948. First
and foremost, his energies were spent on the establishment of new
Jewish settlements in the areas emptied of Arabs. Weitz saw in the
settlement effort, before and during 1948, the primary tool for
entrenching and consolidating the Jewish people's hold on the Land
of Israel and of determining its borders. As well, settlement would
finalize the Palestinian exodus, assuring as nothing else could that
the refugees could never return. He wrote:

The abandoned villages, whether destroyed or standing whole, are a
vacuum . . . which threatens to become a major obstacle to solving the
Arab refugee problem [by way of resettling them abroad]. So long as [the
vacuum] exists, there will also exist pressure to fill it up by a return of [the
original Arab] inhabitants . . . The danger of this vacuum must be removed
by settling Jews [in it].[105]

Weitz's settlement plans and efforts followed hard on the heels of
the IDF's conquering columns. In May, he pressed for the swift
expansion of Jewish settlement in Western Galilee, an area left out
of the Jewish state by the UN Partition Plan of November 1947 and
conquered by the Haganah in the days following the declaration of
independence of the State of Israel. Weitz's attachment to this
fertile strip along the Mediterranean was deep and anguished: in
1946 his eldest son, Yehiam, was shot dead by Arabs near the Az-
Zib (Achziv) bridge there during a Palmah raid. Kibbutz Yehiam,
named in his memory, was set up in the area. The kibbutzim of
Western Galilee, and especially Yehiam, were under Arab siege
from March until the Haganah conquest of the area; their stand as
embattled islands in an Arab sea riveted the Yishuv's attention.

Within weeks of the conquest, the Western Galilee's Arab villages
had been levelled and preparations were well advanced to establish

[104] CZA, KKL 10, protocols of the JNF board of directors meeting of 4 Jan. 1949.
[105] Weitz, iii, pp. 339–40, entry for 11 Sept. 1948.

new settlements: Kibbutz Sa'ar was set up in July 1948; Kibbutz Ga'aton in October. In July, a day after its conquest, Weitz toured the Lydda (Lod) airport area, mentally establishing settlements here and there on the rich 'brown-black-red' soil.[106] The following month, three settlements went up on the abandoned lands of Wilhelma—Nehalim, Atarot (later renamed Bnei Atarot), and Be'erot Yitzhak.

Weitz fluctuated between joy and despair as periods of inactivity seemingly followed each conquest and as the nation's leaders debated a succession of vast settlements plans. 'No movement, the engine is too heavy and the drivers are many', he lamented. It was not enough that the executives of the settlement bodies—Weitz himself, Hartzfeld of the Agricultural Centre, and Yehuda Horin, director of the Jewish Agency's settlement department—wanted new settlements. 'It is necessary that Ben-Gurion will truly want, and that [Finance Minister] Kaplan will want . . . I am in bad spirits', he wrote in June.[107]

Towards the end of the month, at Weitz's urging, the JNF directorate approved a settlement plan involving nineteen new settlements to be set up immediately in the Jerusalem corridor, the Beit Shean valley, and the lower Galilee.[108]

At a meeting with Ben-Gurion in Tel Aviv, attended also by Shkolnik, Weitz, on 23 July, urged that conquests be succeeded immediately by new settlements so that the 'military victories should be translated into political achivements'. But should new settlements also be established outside the areas earmarked in the 1947 partition resolution for the Jewish state and / or on Arab-owned lands?

On the first point, it was agreed that new settlements should go up in the Ramle–Lydda and Jerusalem corridor areas (both lying outside the partition boundaries and strategically straddling the road to Jerusalem). But Weitz noted that Ben-Gurion and Shkolnik, who spoke of '10–12' new settlements beyond the partition boundaries, carefully avoided responding to the question of settling on Arab-owned lands—though plans to settle the Jerusalem corridor implied

---

[106] Ibid., pp. 314–15, entry for 13 July 1948.
[107] Ibid., p. 302, entry for 14 June 1948.
[108] CZA, KKL 10, protocols of the JNF board of directors meeting, 24 June 1948.

as much, as little of the land there was Jewish-owned.[109] Through
August and September, Weitz pressed doggedly for immediate settlement of western Galilee, the central Galilee, and the corridor,
slowly overcoming resistance from the Mapam ministers and
Granovsky, who initially opposed Jewish settlement on Arab-
owned lands. Weitz in this was strongly supported by the defence
establishment. The Defence Ministry in August proposed establishing sixty-one settlements, mostly along the new borders. The settlement institutions—the JNF, Jewish Agency, and Agriculture Centre
settlement departments—preferred a more modest plan, involving
thirty-two settlements, only five of which were to be within the 1947
partition borders and about half on Arab-owned land.

Speaking before the Ministerial Committee for Abandoned
Property, Weitz stressed that the Arab-owned lands proposed for
Jewish settlement were empty of Arabs. But Agriculture Minister
Zisling, from Mapam, persuaded the committee to accept the
caveat that some of the lands of each abandoned village to be settled
by Jews should be set aside for the possible return of the original
inhabitants, political developments permitting.[110] The Mapam
caveat was to remain a formality, and a short-lived one at that.
Hundreds of new settlements had to be established 'on the Arab
lands which were being depopulated', Weitz told the JNF directorate
in September.[111] By December, the idea of setting aside land for
returning Arabs was dead. On 18 December Weitz asked Ben-
Gurion, who was on vacation in Tiberias, whether such land still
had to be set aside. Ben-Gurion replied: 'Not [in sites] along the
border, and in each village we'll take all [the land], in accordance
with our settlement needs, and the Arab will not return.'[112] In any
case, the concept of 'surpluses' had been a fiction from the start.
Each new settlement during the second half of 1948 was allocated by
the settlement bodies as much land as it needed and could handle
without attention being paid to the Mapam-initiated caveat.

[109] Weitz, iii, p. 319, entry for 23 July 1948.
[110] ISA, FM 2564/13, protocol of Meeting of the Cabinet Committee for Abandoned
Property, 20 Aug. 1948.
[111] Weitz, iii, p. 343, entry for 20 Sept. 1948.
[112] Ibid., p. 366, entry for 18 Dec. 1948.

THE SECOND TRANSFER COMMITTEE

While dealing with renewed land purchase from Arabs and the settlement by Jews of successive abandoned sites, Weitz had never lost sight of his original goal—the appointment of an executive body, preferably with himself at its head, to oversee a policy of transfer. The appointment of Danin in mid-July to be the Foreign Ministry's representative in such a body, once established, reinforced Weitz's hopes that it would yet come into being. Weitz met Ben-Gurion and Shertok on 25 July to press for a decision on transfer policy. He achieved what appeared to be movement. Ben-Gurion empowered Shertok to supervise the implementation of the 'Transfer Policy', once again displaying both his desire that the matter be expedited and his wish that he himself not be directly involved. Ben-Gurion failed to record the meeting in his diary.[113]

But Shertok, perhaps busy with other matters (Bernadotte), perhaps waiting for a formal letter of authorization from Ben-Gurion, dragged his feet. Weitz lamented: 'Though the Foreign Ministry has set a line . . . it has not yet [begun] to act on it. A special apparatus is needed which will act and activate [others].'[114] The following day Weitz wrote to Shertok to remind the Foreign Minister of the decisions taken at the 25 July meeting with Ben-Gurion: 'You told me that from now on the initiative in the matter of the 'transfer' would be in your hands and that you would immediately call for consultations and action . . . If I am not mistaken, the consultation has not yet been called and the concrete, practical implementation has not yet been arranged.' Weitz warned that, meanwhile, the whole matter was up in the air, with everybody doing in his own domain as he saw fit. Weitz called for setting up the special apparatus, which would also act to convince the Arabs through propaganda that there would be no return. He asked Shertok not to be angry with his importuning, explaining that he regarded 'the realization of the transfer' as 'the crown of our victory' in the war.[115]

Shertok called in his two chief Middle East advisers, Danin and

[113] The existence and upshot of this meeting are referred to in a letter from Y. Weitz to M. Shertok on 4 Aug. 1948, ISA, FM 2564/20.

[114] CZA, A-246/13, p. 2,456, entry for 3 Aug. 1948.

[115] ISA, FM 2564/20, Y. Weitz to M. Shertok, 4 Aug. 1948.

Ya'akov Shimoni, caretaker director of the ministry's Middle East
Affairs Department, and within a day the two formulated a one-
page memorandum entitled 'Proposals for Immediate Actions
Connected with the Prevention of the Return of the Arab Refugees
to Israel Before A Definitive Policy Regarding the Problem is
Determined'. It was submitted to Shertok on 5 August. The eight-
point document proposed: the collection of data about the present
whereabouts of the refugees and their numbers; listing their
abandoned property; 'A fresh examination of the military and
defence considerations necessitating the destruction of Arab buildings and assistance, in advice and funding, in implementing this
destruction'; collection of information about 'the wandering, flight
or expulsion' of other populations around the world ('Greece–
Turkey . . . Pakistan–Hindustan . . .'); formulating plans for the
resettlement of the refugees in the Arab states; and the appointment
of a 'team of workers' to carry out 'the aforementioned work'. The
document notes that some JNF employees can be mobilized for this
purpose and proposes; 'For the organization of the operation and its
implementation [we] propose the engineer Zalman Lifshitz and Mr.
Hiram Danin.'[116]

That day, Ezra Danin informed Weitz that Shertok had agreed
'that we begin work aimed at the solution of the Arab problem in
line with our our policy: prevention of their return and formulation
of plans to settle them in Arab countries.' Weitz then wrote to
Zuckerman and Nahmani to begin collecting data on the Arab
exodus,[117] and the following day asked Lifshitz to join the still
'unofficial committee for matters of transfer'. Lifshitz agreed.[118]

Weitz spent the following fortnight amassing details about the
movement of Arabs out of Palestine. On 18 August Ben-Gurion
summoned the country's leading Arab affairs and defence experts
for a full-scale consultation on the refugee problem, the possibility
of a return, and policy towards Israel's Arab minority. The meeting,
in Ben-Gurion's office, was attended by Shertok, Kaplan, Shitrit,
the Directors-General of the Treasury and the Minority Affairs
Ministry, Weitz, Danin and Lifshitz, Shimoni and Reuven Shiloah,

---

[116] ISA, FM 2566/13, memorandum entitled 'Proposals for Immediate Actions . . .',
5 Aug. 1948.
[117] Weitz, iii, p. 324, entry for 5 Aug. 1948.
[118] Ibid., entry for 6 Aug. 1948.

the Foreign Ministry's liaison with the defence and intelligence forces, Major-General Elimelech Avner, head of the Military Government in the Occupied Territories, and others.[119]

Apart from the Cabinet ministers, Weitz and Lifshitz were the chief speakers in the debate, which eventually focused on the necessity for barring a refugee return, a policy on which there was complete unanimity. Weitz outlined the dimensions of the refugee problem: 286 villages had been 'evacuated', 179 of these within the Jewish state's partition plan borders. In all, some three million dunams of land had been abandoned by the Arabs. He explained what must be done, closely following the outline-scheme the Transfer Committee had submitted to Ben-Gurion in early June. Some villages had to be destroyed, others settled. What could be bought, should be. Abandoned fields should be cultivated and propaganda begun among the Arabs explaining that the exile was final; there would be no return. He urged that a body be appointed to study the possibilities of resettling the refugees permanently abroad and to prepare 'a plan for the transfer of the Arabs and their resettlement [abroad]'.

When Minority Affairs Minister Shitrit commented that the decision whether to allow a return was not entirely in Israeli hands, Ben-Gurion cut him short with: 'We are proceeding from the assumption of how to assist those who will not return—as many as they will be (and for us it is preferable that they are as many as possible)—to settle abroad.' Shimoni supported Weitz's call for 'an apparatus to investigate, and settle refugees abroad' and subsequently saw in the 'advisory gathering' in Ben-Gurion's office a major step towards getting a transfer committee at last officially endorsed and appointed. The participants, Shimoni wrote the next day to the Paris-based director of his department, Elias Sasson, were unanimous in opposing a return and about doing everything possible to prevent it. Finance Minister Kaplan and his Director-General, David Horowitz, however, 'were more conservative and careful about the means' to be employed 'with regard to Arab property', Shimoni noted.

[119] An abbreviated, four-page protocol of the meeting, written by Ya'akov Shimoni, is in FM 2444/19. It is the best record kept of the meeting. Long entries on the discussion are also to be found in DBG, ii, pp. 652–4 and Weitz, iii, p. 331. In his letter to E. Sasson of 19 Aug. 1948, Shimoni evaluated the previous day's meeting (in ISA, FM 2570/11).

The Cabinet formally approved the appointment of the Transfer Committee during the last week of August, and on 29 August the Cabinet Secretariat informed Weitz that Ben-Gurion had appointed him, Danin, and Lifshitz 'as a committee which must submit to him a proposal about possibilities of settling the Arabs of the Land of Israel in the Arab states'.[120] These terms of reference represented a considerable constriction of Weitz's original conception of the body he had hoped to head. The whole sphere of praxis within the borders of the state, aimed at precipitating further Arab flight and preventing a return, was apparently left outside the committee's purview.

On 31 August the committee met with Cabinet Secretary Ze'ev Sharef to discuss terms of reference for the letter of appointment. That evening, Weitz and his wife Ruhama went to the theatre for the first time 'since [their son] Yehiam's death'. They saw *He Fell in the Fields*, a play by Moshe Shamir about the life and death of a Palmah soldier. 'We wanted to see moments from the life of Yehiam and his death near the [Az-Zib] bridge.'[121] About a fortnight after the committee began work, committee member Lifshitz went to Ben-Gurion to unburden himself of a grave concern: only a minority of the refugees, he related, were living in the neighbouring states proper—'75,000 in Transjordan, 5,000 in Iraq, 12–15,000 in Lebanon and 20,000 in Syria.' The bulk of the *fellahs* 'were encamped [in the Gaza Strip, the West Bank, and Upper Galilee] along the front lines and hope to return'. Lifshitz proposed that these refugees be 'intimidated' into moving away from Israel's borders.[122]

The problem of the refugee concentrations within Palestine along the front lines preoccupied the committee through September, especially against the backdrop of the 18 September publication of the Bernadotte Plan, which proposed the inclusion of western Galilee in Israel with the Negev going to Transjordan, and allowing the refugees to return to their homes. After reading the plan, Weitz became deeply troubled. The plan threatened to bring back 'almost all the Arab refugees through the back door', he commented.[123] Weitz was especially worried by the large refugee concentrations

---

[120] Weitz, iii, p. 336, entry for 29–30 Aug. 1948.
[121] CZA, A-246/14, p. 2,473, entry for 31 Aug. 1948.
[122] DBG, iii, p. 683, entry for 13 Sept. 1948.
[123] Weitz, iii, p. 344, entry for 22 Sept. 1948.

along the Majdal (Ashkelon)–Faluja–Hebron Hills line in the northern Negev approaches and in the upper central Galilee.

The committee formally discussed the problem on 23 September, Weitz expressing the fear that 'the solution of the Arab problem . . . might soon come and not in a manner we had desired'. Danin, according to Weitz, said that a solution to the problem of the refugee concentrations lay in intimidation (*hatrada*) which would force those in the Galilee to move off northwards. Alternatively, Danin suggested, Israel could open talks with the Lebanese and Syrian authorities to persuade them to allow the refugees to move into and settle permanently in their countries. He implied that Israel could grease Arab palms to this purpose. The committee resolved to bring the matter to Ben-Gurion.[124]

Weitz first discussed the technicalities of intimidation with one of his field-men and then, on 26 September, went to see Ben-Gurion. What was to be done about the refugee concentrations? the Prime Minister asked. 'Intimidation', was Weitz's response, 'intimidation using all means.'[125] There were some 100,000 refugees in the central Galilee, Weitz said. They would return to their homes in Israel—in line with the Bernadotte Plan—unless intimidated into moving away.

According to Ben-Gurion, who also recorded the conversation in his diaries, Weitz proposed that the intimidation operation in the Galilee be carried out by Israeli Arabs, Syrians, and Lebanese gendarmes. In the south, the refugees should be intimidated 'without end' (*bli sof*), Ben-Gurion recorded Weitz as saying (but without naming him). Weitz proposed that Shiloah—the Foreign Ministry liaison with the IDF—be placed in charge of the operation, assisted by the Weitz–Danin–Lifshitz committee. 'Weitz asked that I issue orders to Shiloah', Ben-Gurion recorded, not saying how he acted in the matter subsequently.[126]

With the renewal of hostilities in mid-October, almost all the Arabs—locals and refugees—in the south fled or were pushed out of Israeli territory into the Egyptian-held Gaza Strip or the Hebron

[124] CZA, A-246/14, p. 2,492, entry for 23 Sept. 1948.
[125] CZA, A-246/14, p. 2,493, entry for 25 Sept. 1948 and Weitz, iii, p. 344, entry for 26 Sept. 1948.
[126] Weitz, iii, p. 344, entry for 26 Sept. 1948 and DBG, iii, p. 721, entry for 26 Sept. 1948.

Hills, held by the Transjordanian Arab Legion. But in the north, in Operation Hiram, the IDF acted in haphazard fashion, here and there expelling and razing villages, in most places leaving the Arab population, including many refugees from areas conquered by the IDF earlier, *in situ*. Weitz was informed by JNF official Moshe Berger on 29 October that the advance into the Arab pocket in upper central Galilee had begun. Weitz commented: 'There is need to arrange that the refugees will flow from there outwards. I wrote a note about this to Yig[ael] Yadin [IDF chief of operations]'.[127]

But Yadin failed to act on Weitz's advice. Following Operation Hiram, Danin toured the newly conquered area and reported: 'There was no plan about [what to do with] them. The different commanders each acted according to their own judgement. At the same time, reprehensible acts were committed here and there by our side.' Danin thought it was essential to plan in advance Israeli behaviour towards the population of the 'Triangle' (the Jenin–Nablus–Tulkarm sector of the area today known as the West Bank) and the Hebron Hills, which seemed to be next on the agenda of conquest, 'We decided to do this at our next meeting', Weitz recorded.[128] The same lack of a predetermined, unified approach, according to Weitz, seemed to misguide Israeli policy towards the Negev's beduins, some of whom the government was allowing to stay on Israeli territory, over Weitz's strong objections.[129]

The committee's terms of appointment at the end of August had been restricted to theoretical evaluation, investigation, and recommendations concerning the fate of Arabs who had already moved out of Israel. But during the autumn of 1948 the committee as a body and its members individually remained active as consultants and lobbyists regarding the possible fate of various Arab communities. The committee acted as a pressure group, albeit of experts, urging the IDF brass and the country's political leadership to expel Arabs and move refugees away from the new, expanding borders. In effect, this marked a reversion by Weitz to his role in the spring and summer of that year. At the same time, the committee took an active part in promoting and advising about 'unofficial' propaganda

[127] CZA, A-246/14, p. 2,511, entry for 29 Oct. 1948.
[128] CZA, A-246/14, p. 2,522, entry for 9 Nov. 1948, and A-246/14, p. 2,526, entry for 18 Nov. 1948.
[129] CZA, A-246/14, p. 2,526, entry for 18 Nov. 1948.

aimed at convincing the Arab states and the refugees that the exodus was a *fait accompli* and that the refugees would not be allowed by Israel to return. (The public official line was that while the refugees would not be allowed to return during the hostilities, the matter was open for negotiation after the war.)[130]

Danin himself defined the committee's tasks as gathering information about various transfers of population in the recent past (Armenians, Greeks, Kurds, Turks, and so on)—'If we succeed in this amassing [of information], let us hope that we will be able to give every respectable nation its list of crimes in this sphere, in order that they may restrain themselves in their criticism of what is being done in this field in Israel', he wrote—listing and evaluating the abandoned property, and formulating development and resettlement plans for the refugees in the Arab states. Danin was writing the first chapter (historic parallels), Lifshitz the second (abandoned property), and Weitz the third (resettlement schemes), with Hiram Danin acting as committee secretary. However, as if by the way, Danin added that the committee from time to time did deal with practical questions as well, such as 'the problem of infiltration [of refugees back to Israel]; the problem of the large [refugee] concentrations near Israel's borders; and the problem of organized counter-propaganda . . .'.[131] The counter-propaganda campaign involved planting reports in Israel's Arabic newspapers and among foreign correspondents that Israel had no intention of allowing the refugees back.[132]

The 'Transfer Committee' laboured over its report through September and most of October. On 21 October Danin went to Ben-Gurion and informed him that the committee's work was almost done. The two discussed possible Israeli actions *vis-à-vis* the 'Triangle' and the Hebron hills, and Ben-Gurion according to Danin said that he expected the Arabs to flee, which Danin took as encouragement from the Prime Minister for the Transfer Committee's practical activities.[133]

[130] FM 2564/1, Y. Shimoni to M. Shertok, 13 Sept. 1948, and ISA, FM 2570/11, E. Danin (Tel Aviv) to E. Sasson (Paris), 22 Sept. 1948.

[131] ISA, FM 2570/11, E. Danin (Tel Aviv) to E. Sasson (Paris), 22 Sept. 1948.

[132] ISA, FM 2570/11, Y. Shimoni (Tel Aviv) to E. Sasson (Paris), 19 Aug. 1948, and Weitz, iii, p. 346, entry for 3 Oct. 1948.

[133] ISA, FM 2570/11, E. Danin (Tel Aviv) to E. Sasson (Paris), 24 Oct. 1948. Danin mistakenly says that his meeting with Ben-Gurion took place on 22 Oct. 1948.

The committee reported on its major findings and recommendations to Ben-Gurion at his office in Tel Aviv on 26 October. The Prime Minister summarized the committee's eight points as follows:

(A) The Arabs themselves are guilty of their flight.

(B) They should not [be allowed to] return, because they will constitute a Fifth Column, they will hold a grudge [against us] in their hearts, their economy [that is, economic infrastructure] has been destroyed and repatriation [of the refugees] will require giant sums, beyond the state's capacity.

(C) The Arabs who remained inside the state should be treated as equal citizens. . .

(D) The Arabs who fled—will be resettled by the Arab governments in Syria, Iraq and Transjordan . . . [and] Lebanon . . .

(E) The resettlement [costs should come out of] the value of the immovable goods [that is, lands, houses abandoned] in the country (after reparations [for war damage to the Yishuv] are deducted), the Arab states will give land, the rest [will come]—[from] the UN and international institutions.

(F) The extrication of the Jews of Iraq and Syria.

(G) What to do if the Arab states refuse to resettle the refugees?

(H) [What to do] if a return [of refugees] will be forced on us? On no account will we accept a return to the border villages, [and] to the cities only a certain percentage (15 per cent of the Jews [that is, of the Jewish population of each city]), only craftsmen and others who can support themselves. . .[134]

According to Weitz, Ben-Gurion endorsed all eight points.[135] The committee recommended that 235 of the 300 abandoned sites be settled. Ben-Gurion, for his part, suggested that the Jewish communities in the Arab countries be excluded from consideration in the report. He also recommended that all the refugees be settled in one country, preferably Iraq, rather than be dispersed among the neighbouring states. He was set against resettlement in Transjordan.[136]

The core of the report was submitted to Foreign Minister Shertok on 5 November. He approved it 'as a first stage' and said that the committee should continue 'to function as a permanent body affiliated

[134] DBG, iii, pp. 776–7, entry for 26 Oct. 1948.
[135] Weitz, iii, p. 349, entry for 26 Oct. 1948.
[136] DBG, iii, pp. 776–7, entry for 26 Oct. 1948.

with the Foreign Ministry' to be consulted on refugee matters.[137] The completed version of the report, entitled 'Regarding the Solution for the Arab Refugees', and covering several hundred pages, was signed by its authors on 25 November and submitted by Lifshitz to Ben-Gurion on 3 December.[138] The report covered the causes of the flight of the Palestine Arabs, the scope and dimensions of the exodus, and the migration of peoples around the world in recent times.

Unlike other recent migrations, wrote the authors, 'the migration of the Arabs of the Land of Israel was not caused by persecution, violence, expulsion. . .'. Indeed, wrote the authors, the Yishuv tried to persuade its Arab neighbours to stay put. Explicit mention is made of the Arabs of Sheikh Muwannis and Abu Kishk, just north of Tel Aviv, who left despite Jewish entreaties that they 'not leave . . . and [the Jews] promised them defence and protection . . . and mutual cooperation . . . if only they refrained from taking part in the war'.[139] Why then did the Arabs leave? It was

a tactic of war on the part of the Arabs who directed the war against the Jews, whose aim was to remove the Arabs from the neighbourhood of the Jews so that they, the attacking Arabs, would have freedom of action in the war, meaning: To attack, bomb and bombard [the Jews] and the Arabs would not be harmed not by [the attacking Arabs] and not by the Jews by way of retaliation.

Weitz–Danin–Lifshitz wrote that the Arabs expected that their exile would last only a few weeks, after which time they would 'return not only to their own homes but also to the houses of the Jews, and [the Arabs] would inherit the possessions'. In order to make the Arab flight comprehensive, its instigators, reinforced by British Mandate officials, highlighted alleged Jewish atrocities committed against occupied Arab communities. 'The Arabs of the Land of Israel obeyed the order of their leaders in the country and in [the Arab] League, and the Arabs became affected by panic' and fled the country.[140]

---

[137] Weitz, iii, p. 352, entry for 5 Nov. 1948.

[138] Ibid., p. 357, entry for 25 Nov. 1948 and p. 363, entry for 3 December 1948. Also DBG, iii, p. 863, entry for 3 Dec. 1948.

[139] ISA, FM 2445/3, the committee's memorandum entitled 'Regarding the Solution for the Arab Refugees', p. 2.

[140] Ibid., p. 3.

Explaining the destruction of the Arab villages which followed, the authors wrote: 'When the battles developed . . . whole areas became battlefields, many Arab villages were damaged in the actual military actions.' Moreover, the Israeli forces carried out 'retribution' against some hostile villages. The fear of retribution 'hastened the flight and abandonment . . . though in most cases there was no compulsion behind this flight . . . because the Arabs could have lifted a white flag . . . and remained in their places . . .'. Neither Israel's citizens nor its army nor its government 'generally resorted to acts of violence, apart from those [customary] in war, against the Arab citizens to force them to abandon their homes and possessions . . .'. In a special appendix, devoted to unravelling the causes of the exodus, the authors add that the flight was 'deliberately organized' by the Arab leaders 'in order to arouse Arab [feeling] of revenge, to artificially create an Arab refugee problem . . . and to prepare the ground for the invasion [of Palestine] by the Arab states who could then appear as saviours of their brother Arabs'.

The authors specifically link the exodus of the Arabs from the coastal plain at the end of March, beginning of April 'to the end of the citrus-picking season . . . At the end of the season, the Arab Higher Committee gave orders to evacuate the Sharon [that is, coastal plain].' The authors bring this information, they emphasize, to negate charges that the exodus was sparked by the Irgun–Lehi massacre at Deir Yassin, west of Jerusalem, which occurred subsequently, on 10 April. The authors add that at least some of the refugees were Arabs who only years or decades before had immigrated to Palestine from neighbouring Arab states and who, in 1948, because of the compulsions of war and through fear, simply took the opportunity to 'return home'.[141]

The report went on to review the state of the refugees, whose total, according to Weitz–Danin–Lifshitz at the end of November 1948, reached some 460,000. Theirs was a miserable existence, 'encamped under trees . . . or in streets in towns'. It was 'no wonder' they wished to return, wrote the committee. But they had nothing to return to—their houses had been destroyed by war or had been filled with 'Israeli refugees', and their herds no longer existed. 'The only asset the Arab refugees can perhaps take into

[141] ISA, FM 2445/3, the committee's memorandum entitled 'Regarding the Solution for the Arab Refugees', 'Appendix 3'.

consideration is the land. The world *perhaps* needs emphasizing as there is room for the government of Israel to expropriate [the land] as the property of the enemy who fought us. So Czechoslovakia did with all the property of the Sudeten Germans.'

On the practical front, the chief question addressed by the report was whether the refugees should be allowed to return. The authors were unanimous and unequivocal: 'By bringing back the refugees . . . a large Fifth Column would be introduced which would act in [the Arab states'] interests.' Among the Jews, in Israel and the Diaspora, 'opinions are divided', states the report. 'The majority stresses that the refugees should not be brought back . . . under any conditions, and a minority thinks that some of them should be allowed back.' This minority, humanitarian arguments aside, wrote the committee, believed that the continued existence of the refugees along Israel's borders would provoke 'continuous warfare . . . especially as the refugees would breed hatred in the Arab countries for Israel'.

But having considered the pros and cons, the committee ruled flatly against a return. The cardinal fact was that the refugees had already 'transferred' out of the country, a solution always regarded as ideal to the Arab problem in the Jewish state. Allowing them back would be an enormous economic burden; would carry the threat of a 'knife in the back' from the returnees; and would cripple Israel's ability to absorb Jewish refugees wishing to come to the Jewish state —'and when we pose one against the other . . . it is clear that [Jewish] refugees take precedence'. The report went on to recommend that the refugees be settled in three 'thinly-populated' Arab countries—Iraq, Syria, and Transjordan—with stress being laid on the first country, the one with no border with Israel. The report went into great detail about the absorptive capacity of each country and recommended precise areas where the refugees could and should be relocated.[142]

SOLVING THE REFUGEE PROBLEM

The Transfer Committee recommendations regarding Israeli initiatives and help in resettling the refugees in Arab countries were

[142] Ibid. The full Weitz–Danin–Lifshitz memorandum is not on file, but its key 'Introduction' and 'Appendix 3' are available.

never implemented. Rather, Israel and the Arab states in the first half of 1949 focused their diplomatic endeavours on the politics of military disengagement; armistice negotiations and agreements between Israel and each Arab state preoccupied the parties. During the spring, a full-scale peace conference, under the auspices of the UN-appointed Palestine Conciliation Commission, convened at Lausanne. The refugee issue was quickly pushed to the fore by the Arab states, who demanded that Israel agree to a general return as a precondition to reaching any general political resolution of the conflict.

The Arab states were uninterested in vast resettlement schemes, despite the assurance of large-scale US aid; the Palestinians were not eager—they still hoped to return to their former homes; and Israel was interested in promoting and assisting such schemes only within the framework of a general settlement of the Israel–Arab conflict. There was also a growing tendency in the Israeli leadership to link the fate of the Palestinian refugees to that of the Jewish communities in the Arab states, who starting in late 1948 began to immigrate to Israel in large numbers, usually leaving most of their property behind. The dispossession of Palestine's Arabs was equated with and nullified by the dispossession of the Arab world's departing Jews. The problem of compensating the Arabs for the abandoned property, already whittled down by analyses, such as that of the Transfer Committee, of what had befallen Palestinian houses, chattels, and even fields, was ultimately narrowed down to a vague need to compensate only for Arab landholdings. Help in resettling refugees hung fire, pending progress in the political negotiation.

The political deadlock during the spring and summer of 1949 epitomized by Lausanne left the Transfer Committee members deeply perturbed. In March 1949 Danin travelled to London to try to promote resettlement schemes for the refugees, especially in Transjordan. 'I see it as a sacred duty to finish what we began regarding the settlement of the refugee problem', he wrote to Weitz from London at the end of April. While humanitarian considerations may have figured in Danin's thinking, his chief motive was that 'so long as the problem of their [re-]settlement abroad is not solved, your fears . . . will be well-anchored in reality', he wrote to Weitz.[143]

---

[143] Institute for Study of Settlement, Weitz Papers, E. Danin (London) to Y. Weitz (Tel Aviv), 26 Apr. 1949.

Weitz's 'fears' were clear, set down in a letter to Foreign Minister Sharett in late May 1949: 'The chain of bitter refugees surrounding us from all sides with hatred and [feelings of] revenge will not give us rest for many years, and it will stand as an obstacle to reaching real peace between us and our neighbours.'[144]

Danin proposed that Teddy Kollek, an aide of Ben-Gurion's, join him in London, together to seek 'partners' in solving the refugee problem by constructively settling them in Arab countries. Danin believed that there existed 'Frenchmen . . . Britishers and oil people, especially Americans', who could be roped in for such an enterprise. 'Money will solve everything', Danin concluded, adding that Sharett had approved his efforts. Sharett's only fear was that resettling the refugees in Transjordan might in the long term prove harmful to 'the bearded one [that is, King Abdullah]'.[145]

The thrust of Weitz's letter to Sharett in late May was an appeal to pay the refugees the compensation owing for their lands, which would ease the refugees' bitterness, enable them to settle properly in the Arab countries, and salve Israel's conscience to boot: 'If we were ready to purchase peace with the blood of beloved sacrifices, shall we avoid purchasing it with money?' he wrote. Weitz urged the government to issue a clear statement confirming Israel's willingness to contribute its share towards resettling the refugees. Such a statement, he proposed, should be coupled with a firm declaration that Israel would not take back any of the refugees. This would serve to extinguish any hope among the refugees of an ultimate return and undermine the ability of the Arab leaders to stoke the fires of a return in the refugees' hearts.[146]

Through April, May, and June 1949, Danin, sometimes assisted by Kollek, laboured to put together a resettlement package for the refugees. At the start of June he wrote to Weitz that British companies —airlines, construction firms, and others—were showing a keen interest in employing the refugees in the Middle East. Employment would mean earnings, and earnings, said Danin, would mean the growth of a new market for British products. The key problem, according to Danin, was whether Israel would 'allot a substantial

[144] ISA, FM 2444/19, Y. Weitz (Haifa) to M. Sharett (Tel Aviv), 27 May 1949.
[145] Institute for Study of Settlement, Weitz Papers, E. Danin (London) to Y. Weitz (Tel Aviv), 26 Apr. 1949.
[146] ISA, FM 2444/19, Y. Weitz (Haifa) to M. Sharett (Tel Aviv), 27 May 1949.

sum' which would set the ball rolling. Danin had written to Tel Aviv requesting that such a sum be allocated.

But if we receive a negative answer, we shall have to return [to Israel] having achieved nothing, and then he [that is, Sharett] will continue his project of resettling the Palestinian Arab refugees in brilliant speeches [and in the speeches of] Abba Eban [Israel's head of mission to the UN], Sasson, Eytan [the Foreign Ministry Director General] and Zalman [Lifshitz], and the Lord will have mercy on His flock [that is, Israel], which is in need of mercy.[147]

In a letter to Sharett, Danin explained how he was going about finding 'partners' for the refugee resettlement scheme. Marcus Sieff, the British businessman, was approaching British firms on his behalf, wrote Danin, asking that they offer the refugees work in the Middle East. Danin wrote that he had told Sieff that Israel, for its part, would give I£10–15 million towards the resettlement project. This sum, Danin said, was 'the final compensation for the . . . assets' left behind by the refugees. Sieff asked Danin to obtain an official statement from the Israeli government stating how much it was willing to contribute before he would make the rounds of his fellow businessmen. Aware of Sieff's strategy and needs, Danin endorsed Sieff's request.[148] To his letter, Danin attached a memorandum 'redrafted by Mr. Marcus Sieff' on 'The Resettlement and Employment of Palestine Arab Refugees in Middle East Development Schemes'. The memorandum asserted that only a small proportion of the refugees would return to Israel. Their upkeep in refugee camps had so far cost £8 million sterling. Some £30–£40 million sterling were needed properly to resettle '400,000' refugees, who could be integrated into existing or planned regional development schemes. The Arab states sorely lacked manpower, and the Arab governments would come to see the refugees as a boon.

The Sieff–Danin memorandum saw the refugees being integrated into irrigation, industrial, and oil projects in the Arab countries. Specific mention was made of the Ghab Valley reclamation scheme in Syria, the Orontes River dam, Euphrates River irrigation schemes (Iraq), and the construction of the Beirut–Damascus highway.[149]

---

[147] Institute for Study of Settlement, Weitz Papers, E. Danin (London) to Y. Weitz (Tel Aviv), 5 June 1949.

[148] ISA, FM 2444/19 E. Danin (London) to M. Sharett (Tel Aviv), 3 June 1949.

[149] ISA, FM 2444/19, for text of Sieff–Danin memorandum.

No concrete Israeli compensation offer was forthcoming, however, and Sieff wrote to Ben-Gurion in early July that while both Shell and Iraq Petroleum executives had expressed a positive interest in his ideas, implementation of the scheme 'entails the admission of some Arabs to Israel and some payment for Arab property . . .'. Sieff also complained that Israel's public relations regarding the refugees could be improved; all that one read in the British press, he wrote, was of Israel turning back or imprisoning refugees trying to infiltrate back to their homes.[150] No movement on compensation was forthcoming from Tel Aviv, and Danin returned dispirited to Israel at the end of summer. Talk of compensation and planned refugee resettlement schemes coupled with regional Middle East development projects came to naught. The foundering of the PCC talks at Lausanne in August at the same time marked the end of any serious Israeli effort to help solve the refugee problem which, as predicted by Weitz and Danin, was to remain at the core of the Middle East conflcit in the following decades.

The activities of Yosef Weitz and his two 'transfer committees', one 'self-appointed' and semi-official, active in late May and June 1948, and the other, official and active from late August to November, form a major and instructive chapter in the story of the Palestinian refugees. The exodus was a tale both of 'voluntary' flights and expulsion, and many fine gradations between; Weitz's activities, more or less covert, cast light on the causation and unfolding of the exodus among the rural Arab communities and on the evolution of Yishuv policy which, in part, brought it about. As the powerful, energetic head of the JNF Lands Department and a member of key inter-departmental committees, Weitz in 1948 played a major role as executive, adviser, and lobbyist in the emigration of Palestine's Arabs, in the takeover of Arab lands and villages, and in the crystallization of opinion and policy in the Yishuv against allowing a return of the refugees.

In the early spring of 1948, Weitz initiated or prompted the expulsion of a number of Arab tenant-farmer communities from JNF-owned land. Subsequently, his focus shifted to large areas, such as the Beit Shean Valley, Western Galilee, and Ramot–

[150] ISA, FM 2444/19, M. Sieff (London) to D. Ben-Gurion (Tel Aviv), 5 July 1949.

Menashe, in which the JNF had long had an 'interest' in terms of eventual intensive Jewish settlement and regional development. Using both his executive powers (and in June, at least, claiming authority from Ben-Gurion), and his wide range of well-placed contacts in the civilian bureaucracies, the settlements, and the IDF, Weitz was instrumental in emptying these areas of their Arab populations before or just after 14 May.

Weitz was a man of integrity, vision, and action. To the settlements, old and newly established, Weitz was the respected voice of the government and the National Institutions; for the government, Weitz was the spokesman of the landed interests and the settlement bodies of the National Institutions.

While immediately active on the local, *ad hoc* level, Weitz quickly saw the need for decision-making and policy implementation on a national level in order to accomplish the 'transfer' which alone could solve the new Jewish state's 'Arab problem'. Towards the end of May, Weitz organized the self-appointed Transfer Committee. As its head until 1 July, when it suspended operations, Weitz was to show the Yishuv the way in translating the Arab exodus into an irreversible *fait accompli*. After gaining the hesitant oral consent of Ben-Gurion and Shertok, Weitz was to oversee a small but systematic and important operation involving the destruction of abandoned Arab villages with the primary aim of barring a return, ever, of the refugees to their homes. At the same time, he pressed for and organized new Jewish settlements on the abandoned lands, rendering their return to Arab possession all but impossible while creating an ever-expanding, powerful, vested-interest group firmly opposed to a return. The allocation of abandoned fields to established Jewish settlements long in need of additional cultivable land, served the same purpose. Lack of formal, public authorization of his committee's operations compelled Weitz at the end of June to suspend its activities, whereupon he turned to other means of assuring Arab departure and non-return, focusing on efforts to buy out Arab landholders. Weitz was to be frustrated in his efforts to become the chief executive of what he called the 'Transfer policy'. Ben-Gurion preferred that what happened should happen away from the full glare of public appointment and endorsement, and was best carried out on the local level by Jewish settlements and IDF units, especially under the cloak of an enveloping fog of battle. When at last official sanction did materialize, it was for a transfer committee of far smaller, more theoretical scope than originally envisaged by Weitz, with a focus on planning the resettlement of the refugees in the Arab countries rather than overseeing their continued departure from Israel or the conversion of their abandoned villages to rubble or into new Jewish settlements.

But quasi-officially, Weitz and his fellow committee members, Ezra Danin and Zalman Lifshitz, continued to dabble in the praxis of population transfer, mostly in a consultative, lobbying capacity. IDF policy in October–November 1948 towards at least some of the newly conquered areas may have owed something to the advice and lobbying of the committee members.

The Transfer Committee in its report of November 1948 formulated the main line and arguments of Israeli propaganda in the following decades. It denied any Israeli culpability or responsibility for the Arab exodus—denied in fact, its own members' roles in various areas and contexts. It also strongly advised against the return of the refugees and proposed that the government play a major role in promoting refugee resettlement in the Arab host countries.

The Weitz committee recommendations on Israeli compensation and promotion of resettlement schemes abroad proved stillborn. On the other hand, its stand on the question of a refugee return was a powerful prop to those in the Cabinet who in any case were opposed to a return, something almost no one in Israel by the start of 1949 was willing to contemplate.

The political deadlock over the refugee problem effectively frustrated all efforts, even by those so inclined, such as Weitz and Danin, to have the Israeli government pay compensation and promote refugee resettlement abroad. The Arabs' insistence on a full-scale return rather than on resettlement among them played into the hands of those in Israel who were in any case unenthusiastic about paying compensation and regarded the refugee problem, henceforward, as an Arab problem to whose solution Israel, impoverished by war, need contribute nothing. By the summer of 1949, the Weitz committee's positive recommendations were dead. But the refugee problem had become, as Teddy Kollek wrote to Foreign Minister Sharett, 'in many ways . . . the central problem of our foreign affairs'.[151] This was also the opinion of Weitz and Danin.

It is difficult to quantify the effects of Weitz's activities in 1948, when the Transfer Committee was only one of a large number of local and national Yishuv bodies dealing with policy towards the Arabs. In terms of organizing, or prompting others to organize, the expulsion of Arabs, Weitz's direct role may have encompassed only a handful of villages. Similarly, his organization and supervision of the destruction of abandoned villages may have accounted for no more than one or two dozen sites. But indirectly, his personal efforts and those of the Transfer Committee, in both its mutations, probably had a powerful influence on the course of events. On the local level, the settlements and IDF units learned from Weitz and his agents what needed to be done and how to do it. On the national level, Weitz and Danin proved persuasive lobbyists for the transfer cause. On both levels, Weitz provided clarity and guidance in a sphere engulfed by confusion and divided counsels; in the anarchy of the times, in a country undergoing revolution, Weitz attempted to point out the direction and fashion a uniformity of purpose. Weitz's standing in the Yishuv gave legitimacy to the course he was promoting.

The Transfer Committee's consistent and inflexible position from May 1948 onwards against allowing a return of refugees carried a great deal of weight in the decision-making councils of the Yishuv. The committee's codification of its opposition to the return and its reasoning in this context certainly helped to crystallize official and, by a process of osmosis, public opposition to a return. But the committee's 'positive' recommendations, regarding compensation and resettlement of the refugees abroad, for a variety of reasons were ignored. Events and policies in 1949 evolved in such a way as to bar Israeli involvement in compensation or resettlement schemes.

---

[151] ISA, FM 2444/19, Teddy Kollek (London) to M. Sharett (Tel Aviv), 10 June 1949 and Y. Weitz (Haifa) to M. Sharett (Tel Aviv), 27 May 1949.

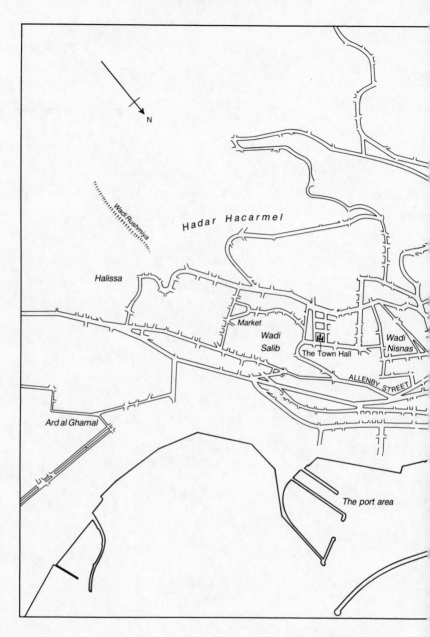

Fig. 3. Haifa in the 1940s

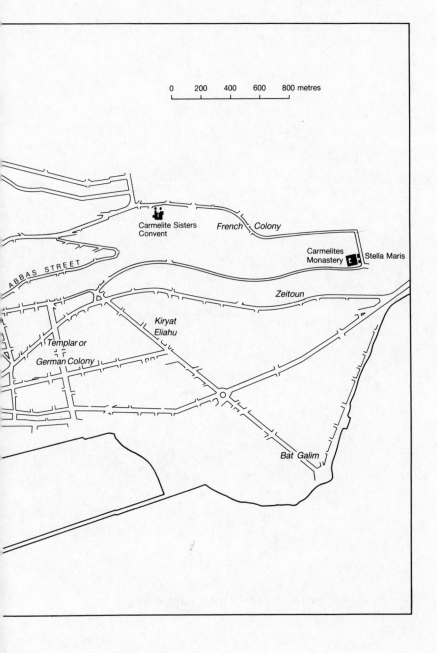

0 200 400 600 800 metres

Carmelite Sisters
Convent

French Colony

Carmelites
Monastery   Stella Maris

ABBAS STREET

Zeitoun

*Kiryat
Eliahu*

Templar or

German Colony

Bat Galim

See overleaf for FIG. 3: Map of Haifa in the 1940s.

5

Haifa's Arabs: Displacement and Concentration, July 1948

Before 1948, Haifa had had a population of 140,000—about half Arab and half Jewish. During the first months of the war, thousands of prosperous and middle-class Arab families fled the city. Tens of thousands more fled Haifa during and immediately after the Haganah conquest of the city's Arab neighbourhoods in April 1948. By July, only some 3,500 of the city's original Arab inhabitants remained.

During that summer, Haifa's remaining Arab inhabitants lived in political limbo. Although the city was part of Israel, by virtue of the United Nations General Assembly Partition Resolution of 29 November 1947 and in consequence of the Haganah's victory of 21–2 April 1948, a lack of clarity prevailed because of the continued presence in the town and its environs—the triangle bounded by Atlit, Ramat David, and Kurdani—of substantial British forces in this last enclave of the Mandate.

During the first week of May, after the battle and mass exodus, the Haganah withdrew the bulk of its troops from the city. The remaining Arab families scattered around the city enjoyed a period of relative calm. But immediately after the completion of the British military's evacuation of Haifa on 1 July, most of the city's remaining Arabs were ordered by the Israeli military governor, Major Rehav'am Amir (Zabledovsky), to pack their belongings, leave their homes, and 'concentrate' in the downtown Wadi Nisnas neighbourhood. A major eviction and moving operation, overseen and carried out by the IDF, got under way, and within ten days almost all the dispersed Arab families were summarily moved and installed in abandoned Arab houses and flats in Wadi Nisnas and along Abbas Street, a second and smaller area of concentration.

Why and how the concentration was carried out will be examined in the following pages. The examination, it is hoped, will shed light on Israeli attitudes and policies toward the Palestinian Arabs during 1948, and on the manner in which the civilian and military authorities of the new-born Israeli state reached and carried out decisions.

PLANS AND PROPOSALS FOR RELOCATION

The idea of concentrating Haifa's Arabs can be traced to a visit to
Haifa by David Ben-Gurion, the Jewish Agency chairman and
Prime Minister-designate on 1–2 May 1948. According to Yosef
Vashitz, a high-ranking member of Mapam and the Arab affairs
correspondent of the party's daily, *al-Hamishmar*, Ben-Gurion had
spoken during that visit of 'his plan regarding the future of the
Arabs of Haifa: (*a*) Their number would not exceed 15,000; (*b*)
two-thirds would be Christians, one-third Moslems; (*c*) all the
Christians would be concentrated in Wadi Nisnas; (*d*) the Moslems
would be concentrated in the Wadi Salib neighbourhood.'[1] According
to Vashitz, who was on the spot, during May and June local Haifa
Haganah / IDF officers tried to persuade Arab families dispersed
around the city to return to, or move to, Wadi Nisnas. 'The plan to
concentrate the Arabs was known to all the commanders,' Vashitz
wrote.[2] But not only the military were interested in concentrating
Haifa's Arabs. In late May, Avraham Ye'eli, the secretary of
Haifa's Arab Affairs Committee, informed his colleagues that he
had submitted a memorandum to Major Amir 'concerning the
delineation of boundaries for Arab residence'. The committee
would 'be able to act' once Amir 'gave his reply'.[3] The Haifa Arab
Affairs Committee had been appointed on 24 April 1948 by Tuvia
Arazi, a senior Haifa Haganah Intelligence Service officer. The
committee was composed of leading Haifa Jews who had Arab
connections and a knowledge of Arab affairs and was set up to look
after the interests of the city's Arab population. It was, in effect, a
subcommittee of the Situation Committee, the supreme authority in
Jewish Haifa until 21–2 April and a major advisory body thereafter.
The Arab Affairs Committee acted as an advisory body and, on

---

[1] HHA–ACP 10.95.10 (4), Vashitz to Secretariat of Mapam Centre (Tel Aviv),
5 July 1948. Ben-Gurion's plan for Haifa was to reverberate in internal Mapam
discussions for months. For example, Mapam's co-leader, Meir Ya'ari, on 26 July
1948, at a meeting of the party's leading military commanders and security officials,
spoke of Mapai's plan to leave '12,000' Arabs in Haifa. HHA, see M. Ya'ari at
'Meeting of Mapam Defence Activists (protocols)', 26 July 1948.

[2] HHA–ACP 10.95.10 (4), 5 July 1948.

[3] ISA, JM, 5667/gimel, 25/2, 'Protocol of the Fifth Meeting of the Arab Affairs
Committee, Haifa', 27 May 1948.

occasion, as an executive body affiliated with and subordinate to the Haganah command in the city.

Ye'eli's interest, at least initially, in 'concentrating' Haifa's Arabs was not motivated by hostility or prejudice. Rather, he linked the need for concentrating the Arabs to a possible return of the Haifa Arab refugees to their former homes—about which, apparently, he was keeping an open mind. He put it this way:

It can be said against us [that is, the Haifa Arab Affairs Committee] that we have still not determined exactly within what boundaries we are willing to allow the Arabs to live in Wadi Nisnas and in Wadi Salib. Of course, the Arabs are not returning [to Haifa] for their own reasons, and from the statements of the people with whom we are in contact it is possible to understand that the reasons [for the Arabs not returning] are not linked [to the problem of the delineation of the Arab pale of residence in Haifa].

Ye'eli went on to explain that Arabs were not returning to Haifa because they feared that the Arab states intended to bomb the city and because they hoped to return on the coat-tails of a victorious Transjordanian Arab Legion. Still others, according to Ye'eli, 'were afraid to return because of the desolation prevailing in the [abandoned Arab districts of the city]'.[4]

But Ye'eli's explanation notwithstanding, the desire of the Jewish civil and military authorities to concentrate Haifa's remaining Arab population in one or two limited areas was in no way connected to a wish to see a return to the city of its tens of thousands of refugees. Rather the contrary was the case. By dint of the Arab evacuation during the last week of April, Haifa, almost overnight, had become a 'Jewish' city. The abandoned Arab neighbourhoods lay open and ready to accommodate the tens of thousands of Jewish immigrants now arriving at Israeli ports; a massive return of Arab refugees to Haifa would have meant many fewer houses for the newcomers.

But it was primarily security considerations that dictated the concentration of Haifa's Arab population. Collectively, Palestinians represented a potential fifth column inside the new-born state that was struggling for its life against invading Arab armies on three fronts. Individually, the Arab populace could produce anti-Israeli

---

[4] Ibid. Ye'eli was being less than frank here when implying that the Arabs, at this time, were not returning to Haifa because they simply did not want to. Many probably did not want to return, but others who did were barred from doing so by the Haganah.

spies, saboteurs, and subversives. The restriction of the 3,000 to 4,000-strong Arab community to one or two clearly defined small areas of habitation would vastly simplify the task of the Israeli security apparatus and substantially decrease the amount of manpower needed by the Haganah / IDF to 'supervise' the community. Haifa's Arabs, moreover, were now defenceless, prey to depredation, intimidation, and assault by such Jewish extremist groups as the IZL and LHI and by Jewish and Arab criminals. Concentrating the community would enable the Haganah / IDF and police to protect them more effectively. It was also felt that isolated Arab families dispersed in mainly Jewish districts or in largely abandoned Arab districts would feel more comfortable if moved to an all-Arab, fully populated neighbourhood. Problems of supplies and employment for the Arabs could also be more easily solved. 'Ghettoization' seemed to afford a simple, quick solution.

The solution, however, could not be effected while British troops remained in control of the last Mandate enclave in Haifa. This enclave was made up of several districts, including the German colony and the Carmelite Monastery of Stella Maris, in which some Arabs lived. The final departure of the British at the end of June 1948 served as a catalyst for Jewish plans to concentrate the remaining Arabs. As Ye'eli reported to a meeting of the Arab Affairs Committee on 1 July, it 'was clear to all of us that with the final departure of the English from Haifa, we would face [the problem of] moving the Arabs [to Wadi Nisnas and, if need be, to part of Wadi Salib]'.

On 1 July, the IDF General Staff ordered the Haifa District Command to carry out the transfer and concentration operation within five days. The army clearly wanted the operation completed before the end of the First Truce, on 8 July, when general hostilities were expected to resume. The order, as transmitted by Major Amir to the Arab Affairs Committee, specified concentrating the Arabs in one area only—Wadi Nisnas. The order caught the committee and Ye'eli by surprise. While he continued to believe that it was 'justified to concentrate the Arabs in one place both from the point of view of the military good and for their own welfare', he resented the speed and abruptness of the order and the fact that 'we were not consulted at all'. This cast doubt on the need for the continued existence of the Arab Affairs Committee. The members would have to make a 'spiritual reckoning', Ye'eli felt. 'One cannot gather up

3,000 persons like eggs. After all, there exist decades-old links between the Jews and many of the Arabs here. In the hastiness of the eviction, how will severe damage to and subversion of these links be avoided?' asked Ye'eli.

Ye'eli agreed that the army had to take into account its own interests—meaning security—'but in such matters it also ought to hear the opinion of others, [and on] how to carry out the thing in a satisfactory manner . . .'. Ye'eli noted that the IDF order asked the Arab Affairs Committee to 'assist in its implementation. But [the committee] was not given any opportunity to give its opinion about the operation it was being asked to carry out.' The army required that the committee provide accurate information about the number of Arabs and their location and about available housing in Wadi Nisnas, and assist in organizing workers and transport for the removal operation. Ye'eli concluded by saying that the committee, especially in light of the opening in Haifa of an office of the Ministry for Minority Affairs, had lost its *raison d'être* and must now weigh whether there was any point in its continued existence.

Vashitz, an Arabist and a committee member who lived in a kibbutz south of Haifa, thought that the committee should continue to exist, if only because the army in Haifa lacked an authoritative adviser on Arab affairs and would fail to take political problems into consideration. As for the eviction order, Vashitz felt that, given the five-day time limit to its implementation, there was 'no possibility of its being carried out in a humane manner', or perhaps at all. It would be preferable, he remarked bitterly, 'to say to the Arabs of Haifa that we regard them as enemies on whom we cannot depend and therefore—we should make them [all] prisoners of war'.

Haifa lawyer Naftali Lifshitz, the Trustee of Enemy Property and a committee member, said that he was 'sorry about the decision [to concentrate the Arabs] which would cause us harm in the wide world'. He saw no point, however, in appealing the order, because it would doubtless do no good. It was better to focus immediately on the task at hand, he said, and organize 'some 50 trucks [and] a lot of inspectors and labourers . . . and act swiftly', making sure that Arab interests would be protected. The meeting concluded with a statement by Major Amir, who was clearly embarrassed by the tone of the committee's criticism of the IDF order. He countered that the committee members had long known of the plan to concentrate the

Arabs. 'It was spoken of more than once', he said. The army had
allotted five days for the operation and 'twelve hours [had] already
gone by since the order was received . . . You will be responsible if
there is a hitch', he told the committee. The 1 July meeting wound
up with the Arab Affairs Committee reluctantly agreeing to assist
the army in the transfer operation in order to make it as clean and
humane as possible.[5]

The leaders of Haifa's Arab community, who included Deputy
Mayor Salah Shihadeh, Victor Khayyat (a millionaire, Spain's
honorary consul in the city, and a US citizen), and Tawfiq Tubi (a
lawyer and communist activist), were summoned to Amir's office
on the evening of 1 July. There they were informed of the plan—to
concentrate the Christians in Wadi Nisnas and the Muslims in Wadi
Salib. Amir led off by stating that he did not know Arabic and would
let Ye'eli explain what was happening. Ye'eli then conveyed the gist
of the IDF order to move the Arabs living in the Carmelite Monastery,
the Germany colony, and other districts to Wadi Nisnas 'to be
implemented by the fifth of (July)'. He asked the Arab representatives to assist in carrying out the operation, so that it could be
completed smoothly. The Arabs immediately protested. Tubi said:
'I see no reason, even military, that necessitates such a step. As I see
it, this is a political rather than military problem. The intention is to
create an Arab ghetto in Haifa. This was not expected by those who
remained here and accepted the existing regime . . . We expected
different treatment . . . We ask that the people [be allowed to]
remain in their places.' Bulus Farah (a Nazareth-born communist
leader) charged: 'This is racism! . . . If there are elements you are
afraid of—take steps to concentrate them, but [do not take steps]
against everybody.'

Amir. I see that you sit here and give me advice. But you were
   invited here in order to hear the Haganah [*sic*, IDF] Command's
   order and to provide assistance! I do not interfere with or take
   part in politics. I only implement orders. During [the past weeks]
   . . . I did all I could in these warlike circumstances to improve the
   conditions of life . . . [But] this is war [and] . . . I must carry out
   orders! I must make sure that this is carried out by the fifth of the
   month . . . I want this work to be carried out by people you

---

[5] ISA, JM 5667/gimel, 25/2, 'Protocol of the Seventh Meeting of the Arab Affairs
Committee, Haifa', 1 July 1948.

appoint and I will give them assistance . . . Though I am a soldier, I also have a heart and it pains me if someone is harmed unnecessarily. I understand and it pains me when a person must be uprooted from a place where he has lived for decades. But what can one do? I don't want to get into an argument because this is not my job. I flatly reject what was said about a 'ghetto' because this is not true. I do not intend to close off the area and seal it with barbed wire . . . No! People will continue their normal lives, but they will be concentrated in one place.

The meeting continued as follows:

Tubi. I take it that the decision is final and cannot be changed. We therefore waive argument, but has the commander considered that there are families that cannot afford the expenses of transfer . . .?

Amir. Tomorrow morning we will deal with the technical matters. I will take care of the vehicles and the benzine, but the people will pay [the other transfer expenses] . . . It is best for you that the matter end quickly, so long as it is in my charge . . .

Shihadeh. There are also Moslems.

Amir. The Moslems will be concentrated in Wadi Salib easily. There are only 174 Moslem families.

Ye'eli. Afterwards [you] will value this operation and see that the concentration is beneficial to you. When the schools open and life proceeds normally, you will learn that all this was only for your benefit. Now you see the thing as a sudden blow. Through the concentration you will also enjoy protection against robbery.[6]

On 3 July, a deputation led by the Greek Catholic Archbishop of Haifa and the Galilee, Georges Hakim, went to see Cyril Marriott, the British consul-general in Haifa, to protest the operation, which had already begun, and to attempt to mobilize British support to stop it.

'I hope it has never been your misfortune to receive a deputation in the circumstances in which I received the Arab deputation this evening', Marriott cabled the Foreign Office. The deputation, which included a number of clerics apart from Hakim, told Marriott that they had come to him because he represented 'his Brittanic Majesty, one of whose titles is Defender of the Faith (most of the

---

[6] Quoted from Tom Segev, *1949 Hayisra'elim Harishonim* (1949 The First Israelis) (Jerusalem: Domino Press, 1984), 69–71. Segev obtained from Amir a copy of a protocol of the 1 July 1948 meeting between Amir and Ye'eli and the Arab notables.

Arabs remaining in Haifa are Christians)' and because 'they now
have nobody else to whom to turn'.[7]

Marriott was very angry with the Jewish authorities. He cabled
London: 'In spite of [Jewish] assurances since the truce . . . that
Arabs who choose to stay would have equal rights with other law-
abiding citizens[,] Jewish military commander of Haifa has ordered
the concentration of Arabs in Wadi Nisnas and Wadi Salib.' He
noted that the operation had begun, with Arab refugee families
encamped on the grounds of the Carmelite Monastery of Stella
Maris being forcibly evicted and the area occupied by Israeli troops.
Marriott complained that one of the consulate's clerks, Hazou, and
a British passport-holder, Zammot, had both been ordered out of
their homes. He subsequently complained to Haifa Mayor Shabbtai
Levy, and these particular evictions were stayed.

Marriott then persuaded Captain Eddy, the representative in Haifa
of the UN mediator for Palestine, to meet the Arab deputation.
Marriott and his American counterpart, Aubrey Lippincott, also
took part in the meeting. Marriott had persuaded the Arabs to
appeal to the UN representative not on 'humanitarian and sentimental
grounds', but on the grounds that the eviction operation was a
breach of the UN-sponsored First Truce then prevailing between
Israel and the Arab states. Eddy, however, responded that he did
not think a truce violation had been committed. Marriott demurred.
'The action of the Jews was a military action with a military purpose
. . . I was satisfied that the Arabs were being moved to the [Wadi
Nisnas] quarter where they would be a fire screen for the Jews.
There was therefore a change in the military situation to the
advantage of the Jews'—hence, a violation of the truce.

Marriott urged London to allocate a ship, the *Ocean Vigour*, for
use by the '1,000' Arabs he believed now wanted to leave Haifa and
embark for Beirut.[8] Marriott, and perhaps the Arab community
leaders as well, were worried about the possibility of Arab air
attacks after the termination of the truce. Wadi Nisnas and Wadi
Salib, he cabled London, 'are near the Jewish quarter and apparently
the intention of the [Jewish] military commander is to surround the
Jewish inhabitants with Arabs so that these parts of the town may be

[7] PRO, FO 371–68568 E8994/4/31, C. Marriott to Foreign Office, 3 July 1948.
[8] PRO, FO 371–68568 E8963/4/31, C. Marriott to Foreign Office, 3 July 1948.

spared attack by air . . ."[9] That same day, in another cable, Marriott elaborated: 'There may be few survivors from the massacre I foresee . . .'.[10] While Marriott's language was characterized as 'somewhat forthright and outspoken' by senior Foreign Office official Lance Thirkell, the 'action of the Jews' in concentrating Haifa's Arabs and the eviction of the families from the Carmelite Monastery was seen as 'going a little too far'. Thirkell felt that the Israeli action should be 'questioned on humanitarian grounds', although he dismissed the idea that the UN mediator could or would do anything about it.[11]

As for Marriott's proposal to supply a British evacuation vessel, Thirkell recommended approaching the Ministry of Transport. A Foreign Office official, J. G. S. Beith, contacted a Transport Ministry official, who ruled that the *Ocean Vigour* was not available as it had previously been used 'for the transport of Jewish illegal immigrants' and was scheduled for refitting. In general, Beith thought that it would be 'far more suitable' for the Arab states to do 'something by way of shipping' rather than Britain, especially as there would be a question of 'where the refugees should go. If we took them off in a British ship, we should have to make arrangements for their reception in some Arab country.' In short, while Marriott's 'concern for these unfortunate Arabs [is] understandable,' said Beith, 'it is not desirable that HMG should involve themselves in this matter. . . . The question of succouring Arab refugees is a matter primarily for the Arab states. Should they wish to charter a British ship . . . we should, I think, agree but that is as far as we should go . . .'.[12]

Harold Beeley, another senior Foreign Office official, more or less concurred. He wrote that it was 'unlikely' that the Jews would agree to an Arab ship entering Haifa harbor to take out refugees and rejected Marriott's request for a British ship.[13] In the event, no

[9] Ibid, and 371–68568 E8994/4/31, C. Marriott to Foreign Office, 3 July 1948. Marriott's 'military geography' makes little sense. Placing Arabs in Wadi Nisnas and Wadi Salib would hardly have 'surrounded' the main Jewish areas of the city, which were to the south and up the slopes of Mount Carmel.

[10] Ibid. Marriott's appreciation of Arab air-power was greatly exaggerated and wholly undeserved. Arab aerial performance was poor in the extreme in the weeks after 15 May and was far from posing a serious threat to any city in Israel.

[11] PRO, FO 371–68568 E8963/4/31, Minute by Lance Thirkell, 6 July 1948.

[12] Ibid., Minute by J. G. S. Beith, 6 July 1948.

[13] Ibid., Minute by H. Beeley, 9 July 1948.

ship was needed. Although a few more Arab families may have left Haifa, the eviction to Wadi Nisnas did not precipitate a new mass exodus. The Haifa Arabs' appeals to Britain and to the United Nations had failed to halt the operation.

The Arabs then pursued a new avenue of appeal—to Haifa's mayor, Shabbtai Levy, who, on 22 April, had issued the dramatic appeal from the Jewish side to the Arabs not to evacuate the city.

On 4 July, Levy cabled the Minister for Minority Affairs, Bechor Shitrit: 'In relation to the concentration of the Arabs in Haifa in one neighbourhood, in Wadi Nisnas alone, it is necessary that you come immediately for a few hours to Haifa.'[14] Shitrit's mandate was the welfare of the country's non-Jewish minority communities. He was informed of the eviction operation belatedly, apparently after it had begun. That day, either just before or just after receiving Levy's cable, Shitrit raised the matter in the Cabinet, submitting a question about 'the isolation of the Arabs in Haifa and Jaffa in special neighbourhoods'. (He also asked that day about a 'mass arrest of Arab villagers' and the destruction of Arab villages).[15] Ben-Gurion, as Defence Minister, responsible for the IDF, apparently ignored the question. The following day, Shitrit cabled Levy that he had raised the matter in Cabinet, but that there was 'no decision yet'.[16] 'No decision' meant failure, as the operation proceeded apace. Mayor Levy seems to have been troubled less by the operation itself than by its restriction of the Arabs mainly to the one area of Wadi Nisnas. Initially, he probably believed that the Arabs were to be moved to the two districts, in line with the original plans.

If Levy's appeal to Shitrit had failed, his intercession at the same time on the local level with Haifa IDF officers seems to have been successful. The eviction order against the wealthy Arab inhabitants of the Abbas quarter overlooking the German Colony was rescinded (probably on 4 July). According to Vashitz, the army initially agreed to allow only the wealthy residents of the area to stay put, but Vashitz intervened and succeeded in persuading Major Amir to adopt an 'egalitarian' approach. Some 50 per cent of the houses and

[14] HMA, 1374, S. Levy (Haifa) to B. Shitrit (Tel Aviv), 4 July 1948.
[15] KMA–AZP 9.9.1., 'Agenda for the Cabinet Meeting, 4 July 1948'.
[16] HMA, 1374, B. Shitrit (Tel Aviv) to S. Levy (Haifa), 5 July 1948.

flats in the Abbas quarter were then allocated to lower-class Arabs, who were chiefly from other districts.[17]

Thus, the plan to concentrate Haifa's remaining Arabs went through three stages: first, concentration in the two quarters of Wadi Nisnas and Wadi Salib, based on religious affiliation; second, concentration only in Wadi Nisnas; and third, concentration mainly in Wadi Nisnas, with an additional smaller area in the nearby Abbas quarter. When precisely Wadi Salib was dropped from the plan is unclear.

'GHETTOIZATION'

The transfer and relocation of the Arabs in and around Haifa— some 720 families—to Wadi Nisnas and Abbas Street began on 2 July and ended on 9 July. It was supervised by the Haifa Arab Affairs Committee, with trucks and troops supplied by the IDF. The Ministry for Minority Affairs, which in theory should have been involved, preferred to stand aside from what it obviously regarded as an unpleasant affair. Moshe Yitah, its newly appointed Haifa representative, made it clear to Ye'eli 'that he did not want to . . . cooperate in the transfer of Arabs as the start of the Ministry of Minority Affairs' activities in [Haifa]'.[18]

On 8 July, with the transfer almost complete, Vashitz and Ye'eli briefed the Arab Affairs Committee, in what the members had decided would be its last meeting; they had voted earlier for dissolution. At the meeting Ye'eli reported that the grounds of the Carmelite Monastery had been completely evacuated, Wadi Carmel 'almost completely', Wadi Jamal and Zaitun 'completely', the German Colony 'in large measure', and the old market-place 'completely'. He believed that the entire evacuation could be completed with a bit of effort within 'a few days'. The Jewish authorities—the Arab Affairs Committee, the IDF, and the municipality, all of which had lent various types of assistance—were also helped in carrying out the operation by an *ad hoc* eight-man committee of Arab notables, selected by each Arab community. It was composed of four Christian priests representing the different

[17] HHA–ACP 10.95.10 (4), Y. Vashitz (Haifa) to Secretariat of Mapam Centre (Tel Aviv), 5 July 1948.
[18] ISA, JM 5667/gimel, 25/2, 'Protocol of the Eighth Meeting of the Arab Affairs Committee (Haifa), 8 July 1948'.

denominations, with the Greek Catholic priest, Butrous Fakhouri, acting as committee chairman, and four secular members, including the two communist activists, Bulus Farah and Tawfiq Tubi, and Haifa's deputy mayor, Salah Shihadeh; Victor Khayyat acted as liaison between the committee and the Jewish authorities. The Arab committee's task was to oversee the removals and to allocate housing in Wadi Nisnas and the Abbas quarter.

Throughout the transfer, the members of the Arab Affairs Committee remained unhappy with the operation but carried it out to the best of their ability in the knowledge that, if they stood aside, the IDF would perform it in a manner that the Arabs would find still more unpleasant. In the middle of the operation, on either 6 or 7 July, the security authorities called a one-day halt and mounted a search for hidden arms among the Arabs. The operation's complexity, the late start, the originally planned five-day span, and the loss of a day in arms searches meant that most of the transfer was carried out in great haste and without even minimal preparation of the flats in Wadi Nisnas and, to a lesser extent, the houses in Abbas Street. 'The flats are filthy, like the streets. There is no water, no electricity, and there are no locks on the doors. There is no time to properly arrange the allocation of flats', Vashitz wrote on 5 July. In mid-operation, there was also a crisis over who was to bear the cost of the operation. 'The army says the Arabs must pay. The Arab [Affairs] Committee has the funds for it but these were impounded by the Minister of Finance and [the committee] is not empowered to spend the money. In effect, every side assumes that someone else will pay.' (The committee's funds derived from the proceeds of sold, abandoned Arab property.)[19]

The Arabs, according to Vashitz, were 'negative' about the whole operation. There was a natural reluctance to be evicted from one's home. Moreover, 'among the simple people, rumours abounded that the Arabs were being concentrated in order to slaughter them more easily, that this [that is, the ghetto] was a prison, etc. The leftists [among the Arabs] opposed [the operation arguing] that this was a political-racist operation and not [a military necessity, and was] designed to create an Arab ghetto in Haifa.'[20] Ye'eli, however,

[19] HHA–ACP 10.95.10 (4), Y. Vashitz (Haifa) to Secretariat of Mapam Centre (Tel Aviv), 5 July 1948.
[20] Ibid.

thought, or said he thought, otherwise. He was 'happy to note', on 8 July, 'that by the decision to settle Arabs also in Abbas Street the cogency of the contention regarding [the creation of] a ghetto was undermined as Abbas Street was not linked geographically at all to Wadi Nisnas'. But there remained the 'danger', according to Ye'eli, that the housing in Wadi Nisnas would be sufficient to accommodate all the Arabs in Haifa, thus making transfers to Abbas Street un-necessary. Ye'eli feared that, in that event, the army would insist that the Arabs be restricted to Wadi Nisnas alone, again raising the spectre of a 'ghetto'. 'I will demand from Rehav'am [Amir] that he fulfil the promise that he gave that he will widen the boundaries [of the Arab "pale of settlement", presumably to include Abbas Street].'[21]

Ye'eli and Vashitz had played leading roles in softening the blow of the transfer and concentration operation. But they disagreed on the operation's purpose and purport. Vashitz was convinced that its roots lay in Ben-Gurion's plans of early May, which were political in motivation and character—'a political act, aimed at creating a ghetto . . . while exploiting an appropriate military moment (*she'at kosher tzva'it*), which partially justifies the operation.'[22] Ye'eli thought that the operation was

essentially military [in nature], and it is also of benefit to the Arabs themselves. For weeks before the Army took the decision, I tried to persuade the Arabs voluntarily to concentrate and, by so doing, many of their affairs [that is, problems] would have sorted themselves out more easily. But they did not agree. Rumours were spread among them that they would be used as hostages; for example, if there was no water in [Jewish] Jerusalem [which depended on water pumped via Transjordanian-held Latrun], we would cut off their water and suchlike.

Ye'eli also remarked that while members of the *ad hoc* Arab transfer committee spoke in contacts with Jews of 'ghettoization', they 'denied this' when speaking of the concentration operation to their Arab constituency.[23]

Initially, the plan had been to settle Christians and Muslims separately. But this plan was scrapped when 'it became clear that there weren't many Moslems [left in Haifa] and that those who had remained were ready to go to Wadi Nisnas', Vashitz said. He

---

[21] ISA, JM 5667/gimel, 25/2, 8 July 1948.
[22] HHA–ACP, 10.95.10 (4), 5 July 1948.
[23] ISA, JM 5667/gimel, 25/2, 8 July 1948.

agreed, as one committee member remarked, that the Christian and Muslim Arabs of Haifa had always lived in largely segregated districts, but added that the Jews 'had no interest in forcing them to stay separate'.[24]

The Arabs offered no physical resistance to the operation and it was completed on 9 July without any outbreak of violence—resistance to the evictions, they felt, would have been fruitless. The operation, all told, cost P£15,000, the cost ultimately being borne by the Haifa Arab Affairs Committee after Lifshitz obtained Treasury approval for the expenditure. The operation was hampered throughout by lack of transport, and one day was lost when the army withdrew its trucks, which were needed for operational duties outside Haifa. 'The heat also caused laxity in the work', Major Amir maintained.

On 8 July, Amir was ordered by the General Staff to complete the eviction and transfer operation that day and was given additional trucks. Apparently, the IDF wanted the operation over before the army embarked on its offensives, which were scheduled to begin the following day, at the end of the First Truce. Amir more or less demanded that the Arab Affairs Committee make a final effort to complete the operation. Ye'eli balked, taking offence at Amir's peremptory tone. 'I do not accept many of Rehav'am's assumptions concerning the order *he* [*my italics, B.M.*] received concerning us as civilians . . . but we will continue to cooperate in the transfer because it is our duty towards the military command, taking account of the war situation in which we are found. Differences of opinion are for other times.'[25]

Vashitz summed up the operation on 5 July by writing that 'there is nothing bad which has not also in it something good'. While the concentration of Haifa's Arab community was, he believed, politically motivated, 'it had renewed [Arab] public life and created a basis for [renewed] internal economic life. Since the start of the concentration [operation], there have been many demands [by Arabs for permits] to open shops, workshops, coffee shops.' He assumed, in mid-operation, that if everything went off smoothly, without violence, 'tempers would [eventually] cool'. 'I believe', Vashitz wrote to the Secretariat of the Mapam Centre, 'that this [operation] is the most important act done with respect to the Arabs in the State of Israel.

[24] ISA, JM 5667/gimel, 25/2, 8 July 1948.   [25] Ibid.

Here it will be decided whether the State of Israel will be a democratic state or a feudal state with medieval customs and Nuremberg laws.'[26]

THE 'SPECIAL CASES'

The July concentration operation, which left most of Haifa's 3,500-strong Arab community in Wadi Nisnas and some dozen families in the Abbas quarter, was to have a sequel. Several dozen Arab families —all 'special cases'—had been passed over in July and had been left in their original homes around the city. Five months later, in December, the Defence Ministry—apparently for security reasons—decided that these Arabs should also be transferred to Wadi Nisnas. But the war was more or less over and the military no longer controlled Haifa. The army wanted the Ministry for Minority Affairs to handle the matter.

On 15 December, two Defence Ministry officers visited Moshe Yitah, the Minority Affairs Ministry representative in Haifa, and informed him that they had received orders 'to concentrate in Wadi Nisnas' the Arab families still dispersed around town—some '110–120', they said. Yitah asked the officers to return with an accurate, detailed list. They returned with a list of thirty-one families. According to Yitah, the list included families already living in Wadi Nisnas, families living in Wadi Salib (an area, he commented, originally earmarked for Arab residence), families with mixed marriages, and families of Arabs holding foreign citizenship. A few of the families, according to Yitah, were living outside Wadi Nisnas and Abbas Street by special individual dispensation of various Jewish institutions. Yitah promptly referred the matter to his superior, Minister for Minority Affairs Bechor Shitrit, who apparently had not been told of the 'special cases' by the Defence Ministry.[27]

Shitrit raised the question of moving these families at the 17 December meeting of the Ministerial Committee for Abandoned Property, a Cabinet body composed of the ministers of defence, foreign affairs, minority, finance, agriculture, and justice. This committee was responsible for the fate of all property of Arab refugees and, in some cases, for the fate of the refugees themselves.

[26] HHA–ACP, 10.95.10 (4), 5 July 1948.
[27] HMA 1374, M. Yitah (Haifa) to IDF Haifa District OC (Haifa), 19 Dec. 1948.

Shitrit reported that the Defence Ministry wished to move '120' Arab families. 'The Minority Affairs Minister is opposed to this and asks [that the committee] take a stand on the matter', records the protocol of the meeting.

The committee, chaired by Finance Minister Eliezer Kaplan, backed Shitrit and ruled that the 'Arabs could not be evicted without the matter being brought before [and approved by] the Ministerial Committee on Transfers [of Arabs from Place to Place]'.[28] Shitrit thereupon cabled Yitah that the ministerial committee had ruled that the Arab families could not be evicted and moved summarily by the Defence Ministry and that he should inform the Haifa IDF command to that effect.[29]

Yitah sent Shitrit's cable to the IDF Haifa District OC, and wrote that the security forces must henceforth consult the Minority Affairs Ministry 'before taking a serious decision', that the ministry must receive orders in writing from the district commander's office rather than learn of them orally, and that the current matter must be considered in light of the ministerial committee's decision, meaning that nothing could or should be done without the approval of the Ministerial Committee on Transfers.[30] The matter apparently ended there. Ben-Gurion did not bring up before the Ministerial Committee on Transfers the matter of transferring fewer than three dozen 'mixed' or 'foreign' families, and the dispersed families were left in place.

OPERATION SHIKMONA

The final British evacuation of Haifa at the end of June triggered not only the concentration of the city's remaining Arab population in Wadi Nisnas and the Abbas quarter, but also opened the way to Operation Shikmona—the destruction of parts of abandoned Arab neighbourhoods and buildings that were ruled unsuitable for accommodating new Jewish immigrants. The destruction, in part, followed blueprints of the city planning authority.

[28] ISA, FM 2401/21/aleph, 'Protocol of the Meeting of the Ministerial Committee for Abandoned Property', 17 Dec. 1948.
[29] HMA, 1374, B. Shitrit (Tel Aviv) to M. Yitah (Haifa), 19 Dec. 1948.
[30] Ibid., M. Yitah (Haifa) to IDF Haifa District OC, 19 Dec. 1948.

Previously, in May, the Haifa Arab Affairs Committee had considered the destruction of 'the Arab shack neighbourhoods in Halissa and Ard al-Ramal for reasons of security and health'.[31] After committee members Ye'eli and Aharon Danin, however, had checked the cost of the prospective demolition work, estimated at some P£10,000, it was decided to abandon the operation. The committee was also swayed by the opinion of Yitzhak Gvirtz, a member of the national Arab Affairs Committee in Tel Aviv, who, on a visit to Haifa, told the Haifa committee that its funds were not designed for such a purpose.[32]

The planning for a face-lift of Haifa's Arab districts really began —as with most developments in the embryonic Jewish state—with Ben-Gurion, who on 16 June consulted with Uriel Friedland (later Shilon), the director of the Shemen factory and a member of the Haganah command in Haifa. Friedland said that there was 'need for demolition [of Arab houses] in Haifa for purposes of adjustment and construction, in line with city-planning [requirements] . . .'. Ben-Gurion agreed and 'urged this work forward immediately with the British evacuation'.[33] Operation Shikmona, carried out by IDF engineer units, got under way in various parts of Haifa during the third week of July, just after the start, on 19 July, of the Second Truce between Israel and the Arab states (which freed IDF units, including engineers, from front-line duties).

The demolition of houses was carried out, according to a complaint by Fred Khayyat, one of Haifa's Arab businessmen, 'without prior announcement, legal process, or promise of compensation'. Khayyat described how he had witnessed the demolition of buildings on Allenby Street and in Hamra Square: 'As far as I can see', he wrote, 'the destruction is being carried out to open up streets marked in the Haifa city planning blueprint', but without the usual procedure of legal expropriation and payment, or promise of payment, of appropriate compensation to owners. As Haifa was completely quiet, wrote Khayyat, the demolitions could not be justified on grounds of

[31] ISA, JM 5667/gimel, 25/2, 'Protocol of the Fourth Meeting of the Arab Affairs Committee, Haifa', 20 May 1948.
[32] Ibid. 'Protocol of the Fifth Meeting of the Arab Affairs Committee, Haifa', 27 May 1948.
[33] David Ben-Gurion, *Yoman Hamilhama-Tashah* [the war diary, 1948–9] (Tel Aviv, 1982), 522, entry for 16 June 1948.

security.[34] A stream of similar complaints by Haifa Arab notables to Mayor Levy[35] prompted the mayor to send a sharp letter on 30 July to Minority Affairs Minister Shitrit: 'I want to make it completely clear to you and the Cabinet that Haifa Municipality has no connection with these operations and, indeed, no formal knowledge of them. All we know is [from] complaints from the people and from hearing the echoes of explosions from parts of the city. Moreover, we do not know under whose orders these actions are being carried out and who is responsible for them.' Levy demanded that the situation be 'rectified' because 'a day will arrive when the owners of the property [that is, mostly Arab refugees] will demand compensation for their property and they will direct their suits at the municipality, believing that it was the municipality which was responsible for the development, expropriation and other projects'. The municipality refused 'material or moral responsibility for the demolitions'. Levy concluded his letter by stating that the municipality rejected absolutely any responsibility for finding housing for those whose houses were being destroyed.[36]

The first official notification any civil authority received from the defence establishment about the demolition operations in Haifa was apparently a telephone call from an officer, who identified himself as 'Barlas', to an official in the bureau of the Trustee of Enemy Property in Haifa on 3 August, about a fortnight after the start of Operation Shikmona. Barlas said he was 'in charge of opening King George Street' and warned that in two days the IDF would blow up two houses on Carmel Boulevard. The official from the custodian's office then called city hall, where municipal officials informed him that the municipality 'knew nothing about the operation'. The official—apparently not Naftali Lifshitz, the Trustee of Enemy Property in Haifa—then wrote to the IDF Haifa District Command and to the mayor saying that he did not know what the legal basis was for the destruction of 'state property'. As far as he could see, it

[34] ISA–MAM 298/82, F. Khayyat (Haifa) to Minister for Minority Affairs (Tel Aviv), 25 July 1948.

[35] HMA 556, Elias Koussa to S. Levy (Haifa), 21 July 1948; Basil Laham, Secretary to the Greek Catholic Archbishop of Haifa (Haifa) to S. Levy (Haifa), 1 Aug. 1948; G. Manjikian and W. Sarkissian, Armenian owners of 'Queen Shoes' (Haifa) to S. Levy (Haifa). 27 July 1948.

[36] HMA, 556, S. Levy (Haifa) to Minority Affairs Minister (Tel Aviv), 30 July 1948. There is a copy of this letter in JM, 5667/gimel, 25/3.

was 'illegal'. He complained both about the 'manner' in which he had been informed of the impending demolitions on Carmel Boulevard and about the destruction itself. 'All those responsible for this operation [Shikmona] will be responsible to the Trustee of Enemy Property and the Government of Israel', he concluded.[37]

The continuing demolitions in Haifa through August and Levy's letter of complaint eventually prompted Minority Affairs Minister Shitrit to take action. On 22 August, he asked his colleagues on the Ministerial Committee for Abandoned Property to put the demolitions on the agenda of the committee's next meeting.[38] But Shitrit missed the meeting, which took place on 27 August, and his stand-in, Minority Affairs Ministry Director General Gad Machnes, failed to raise the subject.[39] Shitrit, however, persisted. At the Cabinet meeting of 1 September, he submitted a question, directed at Prime Minister and Defence Minister Ben-Gurion, about 'the destruction of the houses in Wadi Nisnas in Haifa, the eviction of Arab families from their homes in Jaffa', and other matters relating to Israel's minority communities. Ben-Gurion parried with an evasion and a deception. 'The Defence Minister answered that these matters belonged to the official responsible for the occupied territories.' The reference was apparently to General Elimelech Avner, the OC Military Government, which governed the territories Israeli forces had occupied that lay outside the 1947 UN partition boundaries of the Jewish state. But Haifa, where Jewish military rule had been lifted on 3 May, was not 'occupied territory'; it had been included in the Jewish state in the partition plan. In any case, as Minister of Defence, Ben-Gurion was ministerially responsible for and in charge of the Military Government, a branch of the IDF.

[37] HMA 556, an official signing 'in the name of the Trustee of Enemy Property' (Haifa) to IDF Haifa District OC, 3 Aug. 1948. The legal status of the property of Arab refugees was unclear at this time. *De facto*, it had been expropriated by the State of Israel—and much of it allocated, outright or on lease, to new immigrants, Jewish settlements, government ministries, and IDF units. *De jure*, a Custodian of Abandoned Property, later called the Custodian for Absentees Property, was appointed a few weeks earlier, in July. His task was to oversee and protect the property (houses, land, movable goods), and to oversee its temporary or permanent distribution and sale, and the funds gained from these actions.
[38] ISA, JM 5667/gimel, 25/3, B. Shitrit (Tel Aviv) to the ministers of defence, justice, finance, etc. (Tel Aviv), 22 Aug. 1948.
[39] ISA, FM 2564/13, 'Protocol of the Meeting of the Ministerial Committee for Abandoned Property', 27 Aug. 1948.

The Cabinet's agendas in 1948 were always full; the ministers never managed to cover all the topics listed for each session. Ben-Gurion could usually parry or defer answers to questions on sensitive subjects, especially when presented by a 'weak' minister, as Shitrit was, and when 'buried' inside a whole cluster of queries, as were Shitrit's about the Haifa demolitions.[40]

Shitrit had perhaps not received a satisfactory answer from Ben-Gurion on 1 September, but Haifa's Mayor Levy was at last officially informed on 8 September about Operation Shikmona, and apparently persuaded that all was for the best. On that day, Minister of Interior Yitzhak Gruenbaum visited Haifa and, in a private conversation in the mayor's office, gave justifications for the demolition operation. The gist of the explanation was that because the demolitions were being carried out by the military they could be justified as a military operation and, therefore, did not require compensation for property owners, as would have been the case had the normal city planning expropriation / demolition procedures been adhered to. A great deal of money—of which the State of Israel was short—had thus been saved.

Following the meeting, Mayor Levy wrote to Gruenbaum:

This large-scale operation, which is decisively changing the face of the city, is no doubt of great significance in bringing progress to the city, in accordance with the city building plans which have been prepared over the years. This courageous operation, carried out by the Army by order of the government, will save a great deal of money which the city would have had to pay in the form of compensation for the demolished buildings, had it attempted to carry out the operation by its own [usual] means.

But Levy still had a problem—one of the reasons for his letter—which was that the city urgently needed to clear away the rubble from the demolition sites. (The other reason, apparently, was the Cabinet's wish to obtain Levy's stamp of approval for Operation Shikmona.) Levy said he needed P£10,000 for this and asked Gruenbaum to supply the funds. To illustrate the problem, Levy wrote: 'The destruction of the Old City alone involves more than 300,000 cubic metres of waste . . .'.[41]

If Levy was at last convinced of the propriety and worth of

---

[40] KMA–AZP 9.9.1, 'Decisions of the Provisional Government', 1 Sept. 1948.
[41] HMA, 556, S. Levy (Haifa) to Minister of Interior (Tel Aviv), 8 Sept. 1948.

Operation Shikmona, the local Trustee of Enemy Property was not. At the end of September, Lifshitz against wrote to Levy and the IDF Haifa District Command that to the best of his knowledge, the demolitions—apparently still going on—were illegal and were being carried out without due process. He again condemned the operation and asserted that those responsible would eventually have to explain their actions to the government.[42]

But Lifshitz's protests were ineffectual. The operation, carried out by the IDF on Ben-Gurion's orders, had neatly and cheaply 'rearranged' Haifa, exploiting the background of war, a cowed Arab population unable to protest effectively, and the absence of most of the property owners.

CONCLUSION

The events in Haifa in July 1948 were a sequel to the fall of the Arab neighbourhoods and the evacuation of the vast majority of the city's Arab inhabitants in April, three months earlier. They illuminate first the nature of the Israeli decision-making processes concerning the Palestinian Arabs during the 1948 war; second, the basic divisions within the Jewish community—both between the military and the civilian bureaucracies and among the civilian officials— concerning policy toward the Arabs; and third, the events illustrate the pattern of Israel's treatment of its Arab minority during the war.

All wars are characterized by confusion, and 1948 was probably more confused than most, mixing, as it did, various aspects of guerrilla, civil, and conventional warfare. And for the Yishuv, the war's complexity was compounded by the simultaneous necessity of setting up and operating the machinery of modern government. A great deal of confusion reigned in policy-making *vis-à-vis* the Palestinian Arabs. Little had been foreseen or planned for, decisions were often taken extemporaneously and not necessarily by the bodies nominally responsible for each area. The fount of all authority was Ben-Gurion, and major decision-making in regard to the Arabs—whatever the trappings or dictates of normal democratic government—was in the hands of the defence establishment. A war for survival was on.

---

[42] Ibid., Trustee of Enemy Property (Haifa) to the Mayor, IDF Haifa District OC, 27 Sept. 1948.

Confusion reigned also in the concentration of Haifa's Arabs in Wadi Nisnas and the Abbas quarter and regarding Operation Shikmona. The Minister for Minority Affairs was bypassed or deceived, the Cabinet plenum ignored as the IDF went its own way and changed the face of Haifa. The city's municipal authorities were shunted aside or, in the case of Operation Shikmona, kept in the dark for weeks and months, the civilian Arab Affairs Committee was given its orders by the IDF rather than consulted in an operation that had clear civil as well as security aspects. In general, in both the concentration and demolition operations, a great divide can be discerned between the new-born state's military and civilian authorities in matters pertaining to the Arab minority population during the war. Decisions were made by the defence establishment, and the civil authorities were either kept in the dark or asked to assist in implementation only after the fact.

An indication that this was a wartime state of affairs and that matters would or might change, at least to a degree, once peace dawned, was afforded by the events of December 1948 (by which time all fighting in the north of the country had ceased), when the IDF tried to 'order' the Minority Affairs Ministry to organize the eviction and transfer of several dozen Arab and half-Arab families. The IDF was rebuffed by Minority Affairs Minister Shitrit after he had obtained the backing of a key Cabinet committee.

The events in Haifa in July 1948 also highlighted the division running through the Jewish civilian population concerning Israel's policy towards the Arabs. There were 'soft-liners' and 'hard-liners' (usually following party lines), with continuous argument and tension between the two groups. The hard-line usually won in 1948, both because it enjoyed majority support and because the background was one of a war for survival. The military argument may not always or wholly have been accepted, or even believed, but this was no time to give doubters their head or to take chances. 'Differences of opinion are for other times', as Ye'eli put it, but moral arguments continually surfaced and were heard, even though they did not usually prevail.

The July events in Haifa illustrate the pattern, in the first months of existence of the State of Israel, of the government's policy toward the Arab minority—military rule and arbitrary measures were the order of the day. The Palestinian Arabs, at least many of them, had fought against the emergence of the Jewish state, and those who remained behind to live in it were usually treated as potential fifth columnists or spies. Their treatment had to be primarily a function of security considerations and needs. The wartime situation, however, was exploited to legitimize and carry out various measures— summary eviction and transfers of population, destruction of property without due process or compensation—whose purport was not necessarily bound up with, or mainly tied to, security considerations. The requirements and considerations of the object —the Arab minority population—were not consulted or allowed to affect decision-making. It was war, and the Arab was the enemy.

6

The Harvest of 1948 and the Creation of the Palestinian Refugee Problem

During the 1947–9 Israeli–Arab war, which pitted two more or less agrarian societies against each other, an intense struggle developed, especially during May–July 1948, over the harvest. The struggle was won by the militarily stronger and better-organized side. While it was peripheral to the outcome of the major engagements of the war and of the war itself, it was important in a variety of ways in the creation of the Palestinian refugee problem and hence had a major bearing on the political outcome of the war.

As with the military and other political aspects of the war, the battle for the crops evolved gradually. In some localities, such as the Hefer Valley, neighbouring Jewish and Arab settlements at least initially adopted a live-and-let-live policy of mutual non-aggression; each allowed the other to cultivate and reap his crops. But in other areas, at first spontaneously, Arab and Jewish farmers-militiamen, caught up in the escalating guerrilla operations, attacked each other and each other's fields; the aim was to harass and inflict casualties, as well as to cause economic damage. By early spring 1948, fear and Jewish military pressure prompted Arab villagers in large numbers to begin moving out of the areas allocated to the Jewish State in the 1947 UN Partition Plan. The burning of fields and attacks on cultivators were among the pressures that led to the exodus. During May 1948, with most of the Arabs having left the Jewish-held areas, the Yishuv organized the harvesting of many of the abandoned Arab fields. Over the summer, in various areas, the often hungry and destitute exiles, playing a cat-and-mouse game with the IDF, infiltrated through Israeli lines back to their fields to salvage some of the crop. Jewish fears that, if allowed to succeed, these temporary, economically motivated forays would be transformed into a permanent refugee 'return' led to hard and fast orders to IDF units along the cease-fire lines to prevent such forays with fire.

Thus, the prevention of Arab refugees cultivating their fields coupled with the harvesting of the abandoned fields by the Jewish settlements became one aspect of the general policy, which crystallized during the summer, of preventing an Arab refugee return. The harvest by the Jewish settlements of the abandoned Arab fields created among the settlements an attachment to the newly acquired lands which, in turn, served to harden attitudes against a return of the refugees and contributed to the emergence of the vociferous and passionate 'agricultural lobby' against allowing a return, ever. This Jewish cultivation of Arab fields and the prevention of Arab harvesting continued through 1948 and early 1949, consolidating both the Jewish hold on the newly acquired lands and the refugeedom of Palestine's Arabs.

The first cases of intereference in the cultivation of fields appear to have occurred in southern Palestine in January 1948. While Arab guerrillas fired on Jewish settlements and traffic, and blew up the water pipeline supplying the relatively isolated Jewish outposts, Haganah retaliation often took the form of attacks on Arab farmers in the fields.[1]

As the Haganah attacks increased, the Arabs of the northern Negev (mostly Beduin) and of the northern Negev approaches became 'greatly worried and considerably impressed by the rumour of Haganah threats that the crop will be destroyed if they do not behave themselves', according to the British Gaza District Commissioner. But the threats, according to the British official, would prove an insufficient deterrent, as they 'will not stop [the Arabs from] taking every opportunity of killing such Jews as give them the chance'.[2]

The Jewish assessment of the situation was that the Arabs might resort to the same tactic. Hence Yosef Weitz, the Jewish National Fund Lands Department director who, as chairman of the Negev Committee, was the Yishuv's *de facto* civil governor of the Jewish settlements in the area, thought that the Jewish harvesting of the Jewish fields should be done 'as swiftly as possible, before the

[1] See e.g. PRO, CO 537–3855, CID Summary of Events, 23 Jan. 1948, Appendix, entry for 22 Jan. 1948, which tells of Fakhri Yunis Awad and Yusuf Darwish, two Arab farmers near Isdud (Ashdod) fired upon in their fields by unknown (presumably Jewish) assailants. Awad was killed and Darwish was slightly injured.

[2] PRO, CO 537–3853, 'Fortnightly Report 16–31 Jan. 1948', District Commissioner, Gaza, 3 Feb. 1948, and 'Fortnightly Report . . . 1–16 Feb. 1948', District Commissioner's Office, Gaza, 16 Feb. 1948.

Arabs' concluded their harvest and felt free to renew or increase their attacks on the Jewish fields. The Arabs' fear for their crops had led, apparently, to a falling off of attacks on the Jews. 'Generally, the *fellahin* of [the] Gaza [area] try not to get involved [in hostilities] against the Jews, as their most vital interest today is: To safeguard the crop of their fields, and to assure the possibility of an unhampered harvest.' The Gaza Arabs, and the northern Negev beduin, were rejecting 'until after the harvest' the importuning of the extremists to attack the Jews, according to a Haganah Intelligence Service (Sha'i) assessment in mid-February 1948. However, the assessment added that while many of the beduin were excusing their inaction against the Jews on the basis of fear for their crops, in fact 'most of them' simply were not interested in joining the *Jihad*.[3]

According to the British District Commissioner, R. J. P. Thorne, who was trying to restrain the Gaza Arabs, the 'moderates' were having a hard time controlling the 'hotheads'. He told the Arab community leaders that Arab attacks on the well-fortified Jewish settlements would be 'badly mauled' and that these 'would give the Jews the excuse they were looking for to burn Arab crops at harvest'. In early March, Ben-Gurion noted briefly that the Negev Brigade OC, Nahum Sarig, had reported that the *fellahin* were afraid that 'the Jews would burn their fields'.[4]

The Haganah retaliatory tactic of attack on Arab farmers and fields was not restricted to the south. In March, the British authorities repeatedly reported such attacks in the Galilee, noting that each community feared the burning by the other of their crops before harvest-time.[5]

---

[3] Yosef Weitz, *Diaries* (Tel Aviv, 1965) iii. 243, entry for 23 Feb. 1948; and Kibbutz Meuhad Archive, Palmah Archive (KMA-PA) (Efal, Israel) 109 aleph/204, '*Yediot Tene* (Haganah Intelligence Service daily reports) 16 Feb. 1948'.

[4] PRO, CO 537–3853, 'Fortnightly Report . . . 16–29 Feb. 1948', District Commissioner's Office, Gaza, 1 Mar. 1948 and David Ben-Gurion (DBG) *Yoman Hamilhama-Tashah* [Ben-Gurion's war diary, 1948–9] (Tel Aviv, 1982) i. 291, entry for 10 Mar. 1948.

[5] See e.g. PRO, WO 275–90, CRAFORCE to 17/21 L, 1 Para Bn, etc. 6 Mar. 1948, which cites Jewish attack on Al Samiriya, in the Beit Shean Valley, as an indication that 'Jews are preventing villagers from working on the lands' and PRO, CO 537–3856, CID Summary of Events, 3 Mar. 1948, Appendix 'B,' entry for 2 Mar. 1948, which cites Jewish attack on farmers from the village of Qabba'a, southwest of Kibbutz Ayelet Hashahar. See also PRO, CO 537–3853, 'Fortnightly Report for the Period ended the 29 Feb. 1948', District Commissioner's Office, Galilee District, Nazareth. Other attacks were cited in Mar. by British observers near Deir al Balah,

It was in the Negev, where the wheat and barley fields ripened earliest, that a general policy of destruction of Arab fields was first considered. The imminence of the harvest prompted the Negev Brigade commanders in early April to request from Palmah Headquarters incendiary mortar bombs with which to set alight Arab fields. The options were discussed at a meeting between Yosef Weitz and the Negev Brigade commanders on 28 April. The brigade's troops could hamper the Arabs from harvesting some 100,000 dunams of crops, or agreement could be reached to allow the Arabs to harvest their crops in exchange for Arab agreement to a general cease-fire in the area. Or, if it was decided against allowing the Arabs to harvest their fields, perhaps the Jewish settlements could be allowed to harvest some of the Arab fields, with the proceeds eventually going to the Arab owners. Weitz himself thought 'that the Arabs should not be allowed to reap their fields.' The matter was clearly political, and no decision was reached. But Weitz persisted, and on 4 May wrote to Negev Brigade OC Sarig, either proposing or instructing that Jewish settlements whose fields had been destroyed by their Arab neighbours be allowed to reap the fields 'of the saboteurs' to the extent of their losses. Weitz took a similar line with regard to compensatory harvesting of Arab fields by Kibbutz Mishmar Hasharon in the coastal plain.[6] Thus the mutual destruction of fields was linked with the start of Jewish cultivation of Arab fields.

In some areas, neighbouring Arab and Jewish settlements concluded *ad hoc* agreements of mutual benefit to enable the harvesting on both sides to proceed unhampered. Arabs put out feelers to reach such agreements in the Hefer Valley and Baqa al Gharibiya areas of the northern coastal plain, as did the refugees from Al

southwest of Gaza, and (by Arabs on Jewish farmers), near Kfar Darom, to the east of Gaza. See PRO, CO 537–3857, CID Summary of Events, 23 Mar. 1948, entry for 21 Mar. 1948, and CO 537–3857, CID Summary of Events, 19 Mar. 1948, Appendix 'B,' entry of 18 Mar. 1948.

[6] *Hativat Hanegev Bama'aracha* (Tel Aviv, undated), 49, communication from Negev Brigade HQ to Palmah HQ, 5 Apr. 1948; Weitz *Diaries* iii, 274, entry for 28 Apr. 1948, and p. 377, Appendix No. 1, the Negev Committee to Nahum Sarig, OC Negev Brigade, 4 May 1948; and Labor Archive (Histadrut-Lavon Institute) 235/IV, 2251 bet, Yosef Weitz to Gad Machnes, 2 May 1948, and Mishmar Hasharon to the Arab Property Department (Committee), 21 Apr. and 29 Apr. 1948.

Kheiriya, east of Tel Aviv.[7] But far more common was mutual destruction, which by May had become widespread. Israel Galili, the head of the Haganah National Staff, briefly referred to it in his report on the military situation to the People's Administration (the Yishuv's pre-state provisional government, also known as the 'Thirteen') on 26 April: 'The arson of fields may start, and we will have to mete out arson for arson.'[8] In early May, as the crops ripened, the first reports came in of Arab depredations against Jewish fields in the centre and north of the country. The Golani Brigade reported an incident near Mahanayim, in the Jordan Valley; the Haganah's English-language radio station, Kol Hamagen Ha'ivri, on 7 May reported an attack on Jewish farmers in the fields of Kibbutz Hama'apil; on 9 May, the Yiftah Brigade reported cases of arson the previous day in the fields of the kibbutzim Dan and Dafna, in the Galilee panhandle.[9]

The intensification of the war in May, highlighted by the Arab invasion and the temporary Jewish shift to the defensive after the 15th of the month, coupled with the ripening of the summer crop, saw an intensification in the battle for the harvest. As with the war itself, so with the harvest—Jewish arms generally prevailed. From the Jewish side, the stress shifted (with the successive conquest of new formerly Arab-populated areas) from destruction of fields to the harvesting of the Arab crop; on the Arab side, the stress shifted from destroying Jewish fields to attempting to reach and reap as much as possible in the fields they had abandoned behind or near the Jewish lines.

The decision in principle to harvest as much as possible of the crop of the abandoned Arab villages was taken during the first days of May, probably by Ben-Gurion in consultation with his political and military advisers. An echo of that decision is to be found in the

---

[7] Referred to in LA 235/IV, 2092, 'Circular No. 10', the Samaria Settlements Bloc Committee, 24 Apr. 1948; KMA–PA 100/MemVavDalet/3–138, '*Yediot Tene* (Haganah Intelligence Service Information)' 24 Apr. 1948; 100/MemVavDalet/3–150, '*Yediot Tene* 6 May 1948'; and KMA–PA 100/MemVavDalet/3–154, '*Yediot Tene* 9 May 1948'.

[8] *Minhelet Ha'am*, (ISA, Jerusalem, 1978), protocol of the meeting of 26 Apr. 1948, p. 22.

[9] *Ilan Vashelah*, ed. Binyamin Etzioni (Tel Aviv, undated), 151; Central Zionist Archives, S25–8918, transcript of broadcast on Kol Hamagen Ha'ivri, 7 May 1948; and KMA–PA 109 Gimel/175, Palmah HQ Daily Report to the Haganah General Staff, 9 May 1948.

deliberations of the Yishuv's 'Arab affairs advisers'—who were members of or affiliated to the Jewish Agency Political Department's Arab Affairs Division and / or the Haganah Intelligence Service—with the coastal plain Arab affairs experts in Netanya on 9 May. Among the decisions reached were: 'the [Jewish] farmers [in the Coastal Plain] should be encouraged not to leave unharvested [Arab] crops and that the harvest constitutes part of the war effort . . .'. On the local operative level, the advisers decided that the fields of 'Ein Ghazal—a village not yet abandoned by its inhabitants whose lands lay near the Jewish settlement of Bat Shlomo—should also be harvested. But as these fields were rock-strewn and unharvestable by combine, it was proposed that the harvest be conducted, on behalf of the Yishuv, by the (Arab) villagers of Khirbet Jisr az Zarka, who had thrown in their lot with the Jews.[10]

During May, the Yishuv's settlement institutions, together with the local settlements bloc committees, organized the harvesting of tens of thousands of dunams of abandoned Arab fields—in the coastal plain, in the Galilee, in the Jerusalem Corridor and in the northern Negev approaches. The Agriculture Ministry together with the Arab Property Department (at first a branch of the Minority Affairs Ministry and, from June, part of the office of the Custodian for Abandoned (later, Absentees) Property), allocated the abandoned Arab fields in each area to local Jewish regional councils or settlements bloc committees, which, in turn, distributed them among the Jewish settlements in their region, according to farming capability and needs. In the northern coastal plain, the kibbutzim Gan Shmuel, Ma'ayan Zvi, Mishmarot, Ma'anit, and Ein Shemer jointly harvested over May and June the fields of Baqa al Gharbiya, Ad Dumeria, Arab Kabara, Ali Bek, Kafr Qari, Faras-Hamdan, and other Arab villages; in the Tel-Hai (Galilee panhandle) district, the kibbutzim Dafna, Dan, Kfar Giladi, and Amir, and private farmers from Rosh Pinna, harvested the crops from the fields of the abandoned Arab villages of Nabi Yusha, Qadas, Al Malikiya and the Hula Valley, and so on. Fields inaccessible to combines and other machinery were usually set alight. The Jewish harvest of the Arab fields was seen as 'crucial to the war effort', in

[10] David Ben-Gurion Archive (Sdeh Boqer), 'Summary of the Meeting of the Arab Affairs Advisors in Netanya,' 9 May 1948.

the phrase of Yitzhak Gvirtz, an Arab affairs adviser and head of the Arab Property Department, and brought in income to the Treasury (as the harvesting settlements had to pay the Treasury a set fee for each dunam reaped). Some of the Jewish settlements regarded the harvest as compensation for their material losses during the hostilities and balked at paying. The harvesting triggered possessiveness, and already during the summer, Jewish settlements appealed to the Agriculture Ministry, the Jewish National Fund, and other land-allocating bodies for permanent leaseholds and possession. The harvest of Arab fields in the early summer of 1948 thus became a major step in the process of Jewish acquisition and expropriation of the abandoned Palestinian Arab lands.[11]

The battle for the harvest became increasingly bitter. In the fighting between Kibbutz Ramat Hakovesh and the neighbouring village of Al Tira in the coastal plain, the abandoned Jewish fields—in no-man's land between the opposing armies—magnetized the Arab villagers. The villagers attempted to collect the harvest and the kibbutzniks planted land-mines in the fields. 'It was a war to the finish', recalled a Jewish participant from the Alexandroni Brigade.[12] Similar local struggles raged elsewhere. On the western side of the Galilee panhandle, Arab *fellahin* stayed clear of the fields in no-man's land near the IDF front lines, and dared venture only into the more distant fields opposite Kibbutz Manara and only at night.'A great fear gripped them', recorded one kibbutz member. But, at the same time, the refugees were threatened by hunger. 'If they do not succeed in reaping their [abandoned] fields, it will be a very serious blow . . . The depression and confusion and bitterness among them is great. May these multiply.'[13] The same situation—and similar desperate attempts to infiltrate into fields near the front lines to reap the abandoned Arab crop—were reported at this time in the Hula

---

[11] LA 235/IV, 2082 bet, 'Circular No. 11', Samaria Settlements Bloc Committee, 24 May 1948, and 2251 bet, 'Things Said at a Meeting of Settlements Bloc Committees in Tel Aviv on 26 May 1948', unsigned but printed on the stationery of the Arab Affairs Ministry (the original name of the Minority Affairs Ministry); ISA, MAM 303/41, 'Report of the Department for Arab Property until the end of May 1948', undated, unsigned (but certainly by the department's director, Gvirtz, from early June); and ISA, AM Aleph/19/Aleph (part 1), 'Summary of the Meeting with the Arab Property Committee in the Tel-Hai District, 5 June 1948', unsigned, undated.

[12] *Hativat Alexandroni* (Tel Aviv: IDF Press), 106.

[13] LA 235/IV, 1630, '*Alei Manara*' (the Kibbutz Manara bulletin), No. 65, 4 June 1948.

Valley along the Syrian border, in the Samaria foothills and in the Gilbo'a area.[14]

One Palestinian Arab, who later became an historian, recalled these infiltration attempts during the summer to salvage the crop from abandoned fields. Elias Shoufani, from Mi'ilya, in the Galilee, was 14 at the time. The village's fields, mostly in the coastal area of Western Galilee, had fallen under Haganah control in May. Food supplies in Mi'ilya, which was conquered only six months later by the IDF, were

> scanty . . . The villagers had no alternative but to attempt to collect the crops from the abandoned fields . . . The lure of food was irresistible . . . The trickle of individuals who began to infiltrate across enemy lines in June became a large scale nightly operation in August. Every day at around five o'clock long columns of farmers led their donkeys to the vicinity of the village watching posts and waited impatiently for the sun to set. With nightfall, they made their way through the curving valleys to the plains. All night long they cropped millet from the fields of Kabri and Zib [two abandoned Arab villages]. Before dawn, they loaded the night's crop on their donkeys and returned home.

Later, some of the villagers stayed in the fields during the days, lying low until nightfall, when they would resume reaping.[15]

The plight of the refugees, which was leading to the infiltrations to harvest abandoned fields, in early June was viewed with growing trepidation by the Yishuv's military and political leaders. This infiltration could lead 'in time to resettlement in the villages, something which could seriously endanger many of our achievements during the first six months of the war', stated one intelligence report.[16] The start of the First Truce, on 11 June, aggravated the problem; the temporary cease-fire, the Jewish leaders feared, might tempt, and enable, refugees to try to cross through the lines back to their fields and villages.

On 5 June, three powerful Yishuv executives—Yosef Weitz of the JNF, Elias Sasson, director of the Foreign Ministry's Middle East Affairs Department, and Ezra Danin, a senior figure in the

---

[14] ISA, FM 2570/6, 'Tsur' to Haganah Intelligence Service, 7 June 1948 and '*Ba Tziburiut Ha'Aravit*' (in the Arab Public), Middle East Affairs Department, Foreign Ministry, 11 June 1948.

[15] Elias Shoufani, 'The Fall of A Village', *Journal of Palestine Studies*, 1/4, 1972.

[16] ISA, FM 2570/6, '*BaTziburiut Ha'Aravit*', Middle East Affairs Department, Foreign Ministry, 11 June 1948.

Haganah's Intelligence Service (soon to be named a special adviser at the Foreign Ministry)—submitted a comprehensive proposal to Ben-Gurion pressing the government to decide against allowing a return of the Arab refugees to their homes. The three executives—who the previous month had set themselves up as the unofficial 'Transfer Committee'—outlined a whole range of steps to ensure that there would be no return of the Palestine Arabs (about 300,000–400,000 thus far) who had fled or been expelled from Israeli-held territory. The committee's second proposal (coming after one calling for the destruction of abandoned Arab villages) was 'to prevent all cultivation of land by [Arabs], including harvesting, collection [of crops], olive-picking . . . also during days of "ceasefire" '.[17]

This lobbying, coupled with the increase of harvest-oriented Arab infiltration and growing pressure by local military commanders for a clear-cut policy, resulted during June in repeated orders from Ben-Gurion and IDF OC Operations (and *de facto* chief of staff) General Yigael Yadin to all units to prevent a return of refugees and harvesting by Arabs of Arab fields 'if necessary by [light weapons] fire'.

The crystallization of Israeli policy in this connection is perhaps illustrated by a series of Negev Brigade communications. The unit stationed at T'kumah on 30 May, reported that Arabs were reaping fields some 600 metres away. 'Are we allowed to shoot?' the unit asked HQ. On 7 June the unit stationed at Kibbutz Be'eri informed HQ that it had 'stopped the Arab harvesting . . . We set alight piles of wheat. Several camels, asses and cows were killed . . .'. On 9 June, the unit at T'kumah informed HQ that it had that day sortied out to set fire to the fields of Abu Barjit. 'Again, the standing crop would not burn [but] the [stacked] piles [of crops] burned nicely', the unit reported. At Mishmar-Hanegev the same day the local Palmah unit attacked nearby Arab granaries.[18]

---

[17] ISA, FM 2564/19, 'Retroactive Transfer, A Scheme for the Solution of the Arab Question in the State of Israel,' by Yosef Weitz, Ezra Danin, and Elias Sasson, undated, presented to Ben-Gurion on 5 June 1948 and Y. Weitz to D. Ben-Gurion, 6 June 1948. For a full description of the activities of Yosef Weitz and the two 'transfer committees' that he chaired, see Essay 4 in the present volume and Benny Morris, 'The Crystallization of Israeli Policy Against A Return of the Arab Refugees, April–December 1948', in *Studies in Zionism*, 6/1, Spring, 1985.

[18] KMA-PA 110/2/tet-40, battalion logbook (Negev Brigade), entry for 30 May 1948; 110/2/tet-50, battalion logbook (Negev Brigade), entry for 7 June 1948; and 110/2/tet-51, battalion logbook (Negev Brigade), entry for 9 June 1948.

On 9 June, the Jezreel Valley District commander, in the name of the IDF General Staff, ordered all local units to prevent 'by force' an Arab return and Arab harvesting in their areas; a similar command apparently went out from the General Staff to all IDF units that day or in the following days. On 10 June, six Arab reapers were shot dead in the fields of Al Muharraqa, in the northern Negev. On 13 June, Arabs reaping their fields south of Kibbutz Gesher were fired on and forced to retreat eastwards across the Jordan River. On 15 June, Golani Brigade troops fired on Arab reapers near Zir'in in the Jezreel Valley and near Ghuweir Abu Shusha, north-east of the Sea of Galilee. Again, on 18 June, Golani troops fired on reapers near Zir'in. Reapers were driven off on 24 June, near Tel Shaykh Qaddumi, in eastern Galilee. Jewish farmers were also attacked, near Kibbutz Mizra, in the Jezreel Valley, on 2 July 1948.[19]

On 13 June, General Yadin issued an order to IDF brigades to 'prevent completely' all Arab reaping in areas conquered by the IDF.[20] Yadin expanded on this order on 19 June: 'Every enemy field in the area of our complete control we must harvest. Every field we are unable to reap—must be destroyed. In any event, the Arabs must be prevented from reaping these fields.' According to the director of the Arab Property Department, Yitzhak Gvirtz, 'the army prodded us to immediately reap [the abandoned Arab fields], or else it would burn the crops, and we barely managed to prevent [the army] from doing so'. Gvirtz was writing about fields east of Tel Aviv, in the lower coastal plain, where the army maintained it could not prevent the nightly infiltration of the area by exiled Arab villagers bent on harvesting their fields because of the ban on shooting in force because of the truce.[21]

The General Staff's orders were 'explained' to the Negev Brigade units by Nahum Sarig on 15 June. Sarig wrote:

---

[19] HHA-ACP 10.95.10 (5), 'Levi,' OC Jezreel Valley District, to area commanders, 9 June 1948; KMA-PA 105/390, 'Sergei' (Nahum Sarig) to Palmah HQ, 11 June 1948; KMA-PA 109 Gimel/215, 216, 'Daily Report, 11 June 1948', Palmah HQ to IDF General Staff (covering 9–11 June 1948); *Ilan VaShelan*, p. 239, for harvesting incidents in the North, 13 June–2 July 1948.

[20] KMA-PA 105-422 and 109 dalet-220, Operations/Yadin to Harel, 7th, Negev brigades etc., 13 June 1948.

[21] KMA-PA 103/44, OC Operations to the brigades, 19 June 1948; and LA 235/IV, 2252, Y. Gvirtz to the Custodian for Abandoned Property, 2 Dec. 1948.

In the previous instructions concerning the Arab harvest, it was stressed that our actions in this field are of a political character and their aim is not the destruction of the crop for its own sake. We must [now] know that this approach has changed fundamentally [following] the start of the Egyptian invasion [of the Negev of 15 May]. The destruction of the Arab crop is a direct injury to the invader, [and] forces him [the invader] to worry about flour for the [local Arab] inhabitants, puts him under pressure of the local population and encumbers his transportation system.

Therefore, 'everything must be done to destroy the Arab crop . . .'.[22] Whether Sarig really believed that these were the main or only reasons for the General Staff / Operations general orders in June to destroy Arab crops and prevent Arab harvesting of crops is unclear. Perhaps in the Negev, more than elsewhere, such considerations played an important part in determining military policy *vis-à-vis* Arab crops and harvesting. But in the centre and north of the country, while such military considerations may also have been relevant, the thrust of the reasoning was political and strategic-political—to prevent a return of refugees for a variety of political and strategic reasons.

What was happening to the Arab fields and farmers was intermittently noted, and criticized, by foreign observers. But—unlike Ben-Gurion's internal Yishuv critics, from Mapam—these observers failed to grasp that these 'local' incidents were part of a national policy and design with a clear strategic-political goal.

The UN observers strung out along the front lines during the First Truce reported a number of incidents of IDF troops setting fire to Arab fields. Paul Cremona, the Special Representative of the UN Secretary General, on 23 June complained to Foreign Minister Shertok that these—he cited cases in Khulda, Deir Muheisin, and two other sites—constituted violations of the terms of the truce. Shertok replied three days later that 'no fire was set to Arab crops by Israeli troops. Crops did indeed take fire, but only by accident, the ignition having been caused by exhaust sparks from passing vehicles . . . and by the carelessness of drivers who threw off [*sic*] lighted cigarettes.'[23]

The problem, as the American chargé d'affaires to Damascus put

[22] Nahum Sarig to the units, 15 June 1948.
[23] ISA, FM 2416/9, Paul Cremona to Moshe Shertok, 23 June 1948 and M. Shertok to P. Cremona, 26 June 1948.

it, was that the 'Israelis refuse to permit refugees [to] return to their fields'. The problem of the Arab harvest, wrote Robert Memminger remained 'unsolved'.[24]

The matter was also brought up in the Israeli Cabinet, by Mapam's Agriculture Minister, Aharon Zisling, who throughout 1948 sniped at Ben-Gurion's policy towards the Palestinian Arabs, which Zisling, reflecting majority thinking in Mapam, thought tending towards expulsion. On 27 June, Zisling asked about rumoured plans to destroy forty abandoned Arab villages and about 'the burning of the standing crop' in various parts of the country. Ben-Gurion apparently did not reply. Two months later, the leaders of the Jewish kibbutzim of Ruhama, Nir-Am, and Dorot in the northern Negev complained that the army had burned the fields of beduin in their area who had been 'neutral' in the hostilities and that the exiled villagers of Huj, who had been 'friendly' towards the Jews, were being prevented from reaping their fields.[25]

These complaints, apparently, had, if any, only a marginal effect on Israeli behaviour. On 24 June a senior IDF liaison officer with the UN, Major Chaim Herzog, reported to his commander, IDF Chief Liaison Officer with the UN Lieutenant-Colonel Baruch Komarov, that 'it was decided to somewhat ease [policy] concerning the harvest of Arab crops in no-man's land'. Herzog's discussion with the UN observers had focused on 'the infiltration of Arabs into our lines in order to reap [and] the burning of the Arab fields by our people . . .'.[26]

Be this as it may, in the main things remained unchanged: Arab crops were either burned or harvested by the Israelis, and Arab farmers were prevented from cultivating their fields—the abandoned ones in the rear of the IDF lines and those in no-man's land and near the Israeli lines. Major Herzog had a rather difficult conversation on the matter on 30 June with a senior UN observer in his sector.

[24] US National Archives (NA), RG 84, Jerusalem Consulate, Classified Records 1948, 800-Syria, Damascus (Memminger) to Secretary of State, 30 June 1948.

[25] KMA, AZP 9/9/1, 'Decisions of the Provisional Government', 27 June 1948; and ISA, FM 2564/9, Arye Farda (Dorot), Yaakov Gavri (Nir-Am), and Eliezer Frisch (Ruhama), to the Prime Minister and Defence Minister (Ben-Gurion), 4 Aug. 1948.

[26] ISA, FM 2426/5, Chaim Herzog to 'Baruch', 'Report on the Activities of the Liaison with the UN Observers in the Jerusalem–Tel Aviv Road Area', 24 June 1948.

Major Maliszewsky asked that the IDF permit the exiled Arabs of Abu Shusha, Deir Muheisin, Beit Jiz, and Beit Susin (in the Latrun area) to be allowed to return to harvest their fields during the cease-fire. Herzog said 'no' and the UN man, according to the Israeli liaison officer, 'proceeded with his usual lecture on the inhumanity of war and the disgraceful conditions under which the poor Arabs were allowed to starve while we would not allow them to reap the harvest'. Herzog replied that 'this was war, and that also in accordance with the truce agreement, we were entitled to reap the Arab harvest in our areas. I was immediately informed in an indignant tone that this was sheer communism.' Herzog ended by saying that a 'concession' by Israel was possible on this point if the Arabs allowed the resumption of operation of the Latrun pumping station, which had traditionally pumped up from the Ras al Ayin spring much of Jerusalem's water supply. Nothing came of this. Herzog also turned down a request from the Father Superior of the Latrun Monastery to allow the Arabs of Beit Jiz and Beit Susin to reap the monastery fields.[27]

On 3 July the Secretary General of the Arab League, Abdul Rahman Azzam (Pasha), formally complained to the UN Mediator for Palestine, Count Folke Bernadotte, about the Israeli prevention of Arabs' harvesting their fields behind Israeli lines. According to British diplomats in Cairo, Azzam had ordered Qawuqji's irregulars to shoot 'out of hand' Jews caught burning Arab fields.[28]

Bernadotte himself seems to have fundamentally misunderstood Israeli policy with regard to the harvest of the Arab fields. His reference to the problem in his interim or progress report of 16 September—which was to be his final report—on the Palestinian situation is misleading. 'The period of the first truce coincided with the ripening of cereal crops in Palestine. Since the front line ran almost entirely through land belonging to Arab cultivators, a great number of fields bearing crops [were] in no-man's land and in the vicinity of and sometimes behind Jewish positions [and this] often led the Jews to react by firing on the harvesters', he wrote. But this

[27] ISA, FM 2426/10, Vivien (Chaim) Herzog to 'Baruch', 'Report on a Meeting with Major Maliszewsky, UN Observer Latrun Area, and the Father Superior, Latrun Monastery, on June 30, 1948.'

[28] ISA, FM 2427/9, Abdul Rahman Azzam to Count Bernadotte, 3 July 1948; and PRO, F0371–68570 (E9283/4/31), British Middle East Office, Cairo, to Foreign Office, 9 July 1948.

'military' explanation held true only for Arab fields near and between IDF front line positions. The problem, as construed from Tel Aviv in mid-summer 1948, was essentially 'political' and policy on harvesting was bound up with decision-making about whether or not in general to allow an Arab refugee return. The decision against a return had crystallized, though not in definitive form, in mid-June, and the decisions against allowing Arab cultivation of abandoned fields and favouring Jewish harvesting of the abandoned Arab crops stemmed from it. Military considerations had little to do with it— except in the wider, strategic sense that a massive return of Arab refugees to their former villages and towns within Israeli-held territory could well strategically undermine the Jewish state.[29]

During the following months, hundreds of thousands of dunams of abandoned Arab fields were reaped by Jewish settlements in the north and centre of the country. From August, much of the abandoned land was formally leased out for periods of up to one year to the Jewish settlements. Along the cease-fire lines and in the Negev, most of which was only conquered by the IDF in October–December, IDF units regularly put Arab fields to the torch—sometimes as retaliatory measures though more often in implementation of a policy designed to prevent an Arab return or, as in the south, as part of the campaign to keep the area clear of potentially hostile Arabs. In early 1949, the government organized the harvesting of some 300,000–400,000 dunams of Arab crops in the Negev and northern Negev approaches.[30]

The prevention of the Arab harvest was also used as a tool of policy by the Israeli authorities *vis-à-vis* those Arabs who had preferred to remain in the Jewish state rather than go into exile. The

---

[29] See Benny Morris, 'Crystallization . . .', for a fuller discussion of this.

[30] See e.g. reports on Third Battalion raids on Arab Negev villages and their fields in September in KMA-PA 120, 12 (22 Sept. 1948) and 18 (29 Sept. 1948); and ISA, FM 2426/13, complaints from 5, 10, and 15 Nov. 1948, from Paul Mohn, the Personal Representative of Ralph Bunche, the Acting UN Mediator for Palestine, to Israel Foreign Ministry Director General Walter Eytan, about IDF prevention of Arab cultivation near the front lines. Mohn wrote, on 10 Nov., following IDF firing on cultivators at Al Midya and Ni'lin, on the central front: 'In view of the plight of the Arab civilians concerned, and in order to save them unnecessary hardships during the winter, UN observers have tried to make necessary local arrangements for the harvest, but they have met with difficulties from the side of the Israeli forces.' He asked Eytan to intervene—which Eytan subsequently did, without much success. On 15 Nov. Mohn warned Eytan that 'public opinion in the world' would not take kindly to Israel's treatment of the Palestinians, in connection with the harvest.

Middle East affairs experts at the Israel Foreign Ministry put it rather bluntly at the end of July, during the Second Truce: 'In many parts of the country the matter of collecting the harvest served as an important card in our hands. It was not without worth that the Druze in the north of the country [who had in April switched allegiance from the Arab to the Jewish side] were allowed to reap their crops while their neighbouring [Muslim] villages were barred from doing so and they are wandering around hungry.' The implication is that the Druze freedom to harvest their fields had been a factor in the retention of Druze loyalty by the Jewish side.[31] The same use was made by the Israeli authorities of the right to harvest in the case of Mahmoud al Mahdi, a Muslim, of Ijzim, in the Coastal Plain.[32]

In general, with many of the still-inhabited Arab villages (mostly in the Galilee) close to the front lines, Arabs were barred by the IDF and security services from cultivating their fields. Part of the problem was technical. The property of those who had fled the country now 'belonged' to the state. In the case of most of the villages—where all the inhabitants were no longer *in situ*—matters were simple; all the land 'belonged' to the state and could be, and was, distributed among Jewish settlements. But in Galilee, many villages remained at least partially inhabited; it was often difficult to distinguish between fields owned by those who had left the country (which now belonged to the state and could not be cultivated by the Arabs) and fields owned by villagers who had stayed put (which remained the possession of the villagers). But more important, the Arabs who had stayed put were placed under Military Government, and their freedom to move freely outside their villages was severely curtailed. The daily movement of villagers back and forth from village to fields (sometimes involving relatively great distances) was regarded by the authorities as a security risk or, at least, problem. The country was still at war with the Arabs; the Israeli Arab minority was seen as a bed of potential saboteurs, spies, and fifth columnists. Hence in most areas of the country, the Military Government and local IDF units found it simplest to forbid *in toto* Arab cultivation of

---

[31] KMA-PA 100/MemVavDalet/1–22, '*BaTziburiut Ha'Aravit*', 20 July 1948.

[32] The case of Mahmoud al-Mahdi took up months of the time of Israeli Foreign Ministry and Defence Ministry officials. See ISA, FM 2570/1 for the protracted correspondence in the autumn and winter of 1948.

fields. At the same time, Jewish settlements began to cultivate fields of Arabs who had remained in the State. This is what happened, for example, to fields belonging to Shafa 'Amr inhabitants in November.[33]

The situation in the Galilee was summarized in April 1949 by the Mapai boss in Haifa, Abba Khoushy, in a speech to the party's Knesset members: in many villages, the Arabs were not being allowed access to their fields; some villages had remained 'without land'; their crops were being pillaged.[34]

Eventually, the Galilee lands were sorted out. Those belonging to absentees reverted permanently to the state, and were mostly distributed among the old and new Jewish settlements. And as the dust of the first Israeli–Arab war settled and the Military Government took efficient control of its wards, fears subsided and the Arab farmers still living in the state were allowed to resume cultivation of their fields.

But during 1948 and the early months of 1949, Israeli policy towards the Arab fields and the harvest of the summer and winter crops had contributed to the creation of the Palestinian refugee problem. The retaliatory policy of early 1948, involving the burning of Arab crops, had undermined rural Arab morale and self-confidence and to some extent impoverished Arab villagers; in some areas there was a lack of flour, or fears of an impending major shortage of flour. It certainly contributed to the readiness of the Palestinians to go into exile during the spring and summer. It is probable that at least some Haganah commanders understood that the arson of the fields would have this effect.

In May and June, a dual policy evolved in the Yishuv—to bar the Arab villagers, now in exile, from returning to harvest their fields and for the Yishuv to reap as much of the Arab crop as Jewish equipment and manpower allowed. What could not be harvested was destroyed. The policy effectively deepened the psychological and physical separation of the Arab *fellah* and tenant-farmer from

[33] ISA, MAM 302/83, Minority Affairs Ministry Haifa Branch office to the Minister for Minority Affairs, 9 Nov. 1948, and A. Hanuki, Agriculture Ministry, to the Minority Affairs Ministry, 28 Nov. 1948.

[34] ISA, Finance Ministry Papers 10/1/9, 'Protocol of the Meeting of the Leadership of the (Mapai Knesset) Faction together with Our Comrades in the Cabinet and Invited Guests on the Problem of Activity Among the (Israeli) Arabs', 26 Apr. 1949.

his lands and home, reinforcing his sense of, and existence as an, exile. In the Negev, still largely in Arab hands, the prevention by fire and sword of Arab harvesting was one direct cause of the Palestinian exodus. At the same time, the massive cultivation of the abandoned Arab fields by the Jewish settlements around the country forged an initial, powerful, and protracted link between these settlements and the newly cultivated fields. This link, within weeks, was to breed a sense of possessiveness which, in turn, led to demands for a permanent transfer of these lands into Jewish hands. The need, after the summer harvest, to sow the fields with winter crops and the state investment in seed-funds consolidated the hold of the state and the settlements on the abandoned lands. As agricultural pioneer and Mapai stalwart Shlomo Lavi put it in late July 1948: 'We have reaped the Arabs' fields and now a new [agricultural] year is upon us.' He proposed that the fields now be ploughed and sowed. 'We must not leave uncultivated fields', he told his Mapai colleagues. 'There is nothing more terrible for a farmer than to see fields of thistles, it is one of the greatest sins.' He suggested that the profit from the harvest could be given to the exiled landowners—if they resettled permanently in the Arab countries. Implicit in the appeal was the idea of permanent expropriation of the abandoned lands.[35]

The Jewish settlements and settlement associations, in which Lavi was a leading figure, during the summer became powerful interest groups militating against a return of the refugees. The day after Lavi's statement at the Mapai meeting, Reuven Cohen, a leading figure in the Kibbutz Meuhad kibbutz movement, one of the mainstays of Mapam, put one of the interest group's arguments very clearly:

The question is how to maintain the [Yishuv's] economy. One of the main things that could save us is extensive cultivation of [our lands] and, in addition, the cultivation of the Arabs' lands. This would mean an addition of about 1,000 dunams for each kibbutz . . . This is a plan and opportunity for additional income for the kibbutzim . . . The function of the [movement] is to press for the establishment of new settlements and to organize the settlement groups.

[35] LPA 23 aleph/48, protocol of the meeting of Merkaz Mapai (the Mapai Center), 24 July 1948, statement by Shlomo Lavi.

Cohen in this connection mentioned the abandoned Arab villages of Al Tira, Tantura, Buteimat, and Al Kafrin, all south or southeast of Haifa.[36]

The position taken by these powerful pressure groups (which provided the backbone of the defence establishment), in turn, hardened the Yishuv leadership's resolve, which effectively crystallized over the second half of 1948, not to allow a return of the refugees and not to give up any of the newly acquired lands.

Through the summer, autumn, and winter of 1948 the Israeli security forces prevented Arab cultivation of abandoned fields in the rear of the IDF lines and near them, and the cultivation of 'unabandoned' lands (those whose owners had stayed in Israel). In large measure the reason for this policy was military; constant movement across the borders by infiltrators invited and heralded trouble for the Jewish state. And such 'temporary' forays into Israel could well lead to resettlement by the refugees on a more permanent footing in the desolate villages, representing a strategic-political (fifth column) threat to the new-born Jewish state. Be that as it may, the policy and measures designed to prevent Arab cultivation of the abandoned fields served to consolidate the separation of the exiles from their lands, driving home the message that there would be no return, which, in turn, further demoralized the exiles and made it that much easier for Israel to keep them out. At the same time, the prevention of cultivation by Arabs still resident in the state and the takeover of most Arab fields by the Jewish settlements cannot but have persuaded some of the exiles that they would face an uncomfortable life if they did indeed return; and it may have persuaded some of those who had so far remained in Israel that life would be more pleasant, after all, in exile. It is possible that some of those who fled before the advancing IDF in October 1948 in the Galilee and Negev were at least in part motivated by the knowledge of what had befallen their cousins who had come under Israeli rule in the months before.

---

[36] KMA 1 bet/8/35, protocol of the meeting of the Kibbutz Meuhad Secretariat, statement by Reuven ("Vinya") Cohen, 25 July 1948.

7

The Case of Abu Ghosh and Beit Naqquba, Al Fureidis and Khirbet Jisr az Zarka in 1948—or Why Four Villages Remained

In the course of December 1947–August 1949, the great majority of the Arabs living within the territory that in the first Israeli–Arab war became Israel went into exile and became refugees. Most of these refugees were created during the bitter fighting of April–June and July 1948. Those who remained behind, becoming Israel's Arab minority, were concentrated almost exclusively in the areas that fell to Israeli forces towards the end of the war, in October–November, or were transferred to Israeli sovereignty through the Israel–Jordan General Armistice Agreement of April 1949, at war's end—the central and upper Galilee, the northern Negev (beduins), and the western foothills of Samaria (Umm al Fahm and Wadi Ara, Taibe, Baqa al Gharbiya, Kafr Kassim).

Nowhere was the clearing of Arab communities out of the Jewish State more thorough and complete than along the new state's two major highways, the Tel Aviv–Haifa road on the Mediterranean coast and the Tel Aviv–Jerusalem road running through what became known as the 'Jerusalem corridor'. The great majority of the Coastal Plain and 'corridor' Arabs fled over March–May 1948—out of fear, Jewish and Arab pressures, and Jewish attack and expulsion.

But there were four notable exceptions: the villages of Abu Ghosh and Beit Naqquba, west of Jerusalem, sitting astride the Tel Aviv–Jerusalem highway, and Al Fureidis and Khirbet Jisr az Zarka (also called 'Arab al Ghawarina), south of Haifa along the main Tel Aviv–Haifa coast road. These villages remained, and have grown and prospered along with Israel itself during the past forty years. Why these villagers were allowed by the Yishuv to stay put and what, if any, was the significance of these exceptions, will be examined in the following pages.

ABU GHOSH AND BEIT NAQQUBA

In late June 1948, during the month-long First Truce between Israel and the invading Arab states, the IDF's Harel Brigade, responsible for the Jerusalem corridor, made clear to the local Jewish settlements and to the national political leadership its desire and intention to expel the villagers of Abu Ghosh and Beit Naqquba before the resumption of hostilities, which all expected (and indeed occurred) on 8–9 July.

Abu Ghosh was regarded by the Yishuv as a 'friendly' (some Arabs might and did say 'collaborationist') village. Traditionally, its relations with neighbouring Kibbutz Kiryat Anavim—during 1948 the site of the HQ of various Harel Brigade units—were excellent. The village had occasionally sold arms to the kibbutz; it regularly provided the Jewish defenders—who included a Palmah company —with information about hostile Arab intentions, armaments, and alignments. During the Mandate, the Arab villagers repeatedly shielded or hid Jewish—both Haganah and LHI—militiamen and terrorists being sought by the British authorities. Throughout the 1936–9 Arab Rebellion in Palestine, Abu Ghosh defied the Arab irregulars (called by the Yishuv *knufiyot* or gangs) and in various ways assisted Kiryat Anavim. 'They worked together with the [British] army against the gangs', according to Ya'akov Lisser, the *mukhtar* of the Kiryat Anavim.[1]

Abu Ghosh adopted the same attitude during the first months of the 1948 war, supplying Kiryat Anavim with 'small quantities of ammunition'. Other Arab villages treated the inhabitants of Abu Ghosh as 'enemies and traitors'.[2] Throughout the first months of the war, the villagers had displayed 'not only loyalty but a friendly attitude towards the Yishuv . . . [Jewish] convoys that were attacked found sanctuary in Abu Ghosh'. Kiryat Anavim requested funds

[1]   ISA, FM 2564/9, 'Memorandum of the Meeting on 6 July 1948 at the Police Ministry, between Mr. B. Shitrit, the Minister of Police and Minority Affairs, and Mr. Ya'akov Lisser, who Represents Kiryat Anavim in the Question of the Arabs of Abu Ghosh', 7 July 1948. But relations between Kiryat Anavim and Abu Ghosh were apparently not quite so consistently halcyon as later myth asserted. One kibbutz old-timer, writing in the settlement's volume recalling 1948, *Al Em Haderekh* (Kiryat Anavim Press, 1949), 5, states that 'we also suffered from Abu Ghosh not a little, though it [generally] displayed towards us good neighbourly relations . . .'.

[2]   ISA, FM 2564/9, 'Memorandum of the Meeting on 6 July 1948 . . .'.

from the Jewish Agency for Abu Ghosh, which was in economic straits, and during the following months supplied the village with provisions.[3]

Relations between Kiryat Anavim and Beit Naqquba—in effect a satellite village of Abu Ghosh—were never quite as good. The kibbutz had always coveted and repeatedly (and unsuccessfully) tried to buy tracts of land from Beit Naqquba. During the Arab Rebellion, Beit Naqquba apparently hosted Arab irregulars passing through and occasionally there had been sniping between the two communities.

During April and May 1948, Haganah units, spearheaded by the Palmah's Harel Brigade, fought bitter battles (operations Nahshon, Yevussi, Maccabi, and Bin-Nun A and B) against the irregulars and the Transjordanian army, the Arab Legion, for possession and dominance of the Tel Aviv–Jerusalem highway, the only lifeline to Jerusalem's besieged 100,000-strong Jewish community. In the course of the fighting, almost all the Arab villages in the corridor were depopulated, either by flight or Israeli expulsion. Abu Ghosh, and smaller Beit Naqquba, remained *in situ*, despite occasional requests by local Palmah commanders for permission to expel or nudge the villagers into at least temporary flight.

In early June pressure built up to expel Abu Ghosh. Yishuv intelligence sources reported that 'people from Abu Ghosh are giving information to the enemy, either willingly or under duress'. The continued presence of the villagers near IDF bases was causing 'demoralization'. Ya'akov Shimoni, of the Foreign Ministry's Middle East Affairs Department, an old Haganah Intelligence Service hand, thought that perhaps the time had come to 'expel [Abu Ghosh], while [maintaining friendship] with them', a position he had long advocated. But he added a caveat that the possibility of using Abu Ghosh for contrary, Yishuv intelligence purposes also needed examining.[4] But a week later, the caveat was removed, and Shimoni took action. On 15 June he met Lisser and another man named 'Segal' and concluded with them that 'on the eve of the renewal of hostilities, they would give forceful advice [to the inhabitants of the village] to pack up and leave. Meanwhile, they

---

[3] CZA S25-9.194, Kibbutz Kiryat Anavim to the Political Department of the Jewish Agency, 29 Apr. 1948.
[4] ISA, FM 2570/6, Shimoni to Eliahu Sasson, 9 June 1948.

would continue to adopt the usual attitude towards [the villagers] and would not say anything [to them].'[5] From the context, incidentally, it would appear that Shimoni had meant departure of Abu Ghosh's population to Arab territory rather than transfer inland, into the interior of the Jewish State.[6]

Meanwhile, more or less simultaneously, the local army command, probably led by Harel Brigade OC Colonel Yosef Tabenkin, also decided that the time had come for expulsion or transfer. 'The military believe that one must not leave inside our area an Arab village like Abu Ghosh with a population of 1,000 and [like] Beit Naqquba with a population of 300 . . . The army wishes to move the Arabs from Abu Ghosh and clean up the place [*lenakot et hamakom*].'[7]

Word of this quickly filtered out to Minority Affairs Minister Bechor Shitrit, who on 30 June raised the matter 'urgently' in the Cabinet. The proposal mooted, apparently, was to transfer the inhabitants of Abu Ghosh (and perhaps also of Beit Naqquba) to Jaffa, from which most of the original Arab population had fled out of fear and under the impact of the Irgun Zva'i Leumi (National Military Organization or 'Irgun') assault, in April and early May. Prompted by Shitrit and the two Mapam ministers, the Cabinet blocked the IDF resolve by deciding that only the 'Ministerial Committee for the Abandoned Area' (soon officially designated the Ministerial Committee for Abandoned Property) could decide on such an expulsion.[8]

But Harel Brigade, especially in view of the imminent renewal of hostilities, still wanted the villagers out. On 6 July, Lisser went to Tel Aviv to meet with Minority Affairs Minister Shitrit. Shitrit, who had apparently initiated the meeting, informed Lisser of the 30 June Cabinet decision that the fate of Abu Ghosh was in the hands of the Ministerial Committee for Abandoned Property. He asked Lisser for Kiryat Anavim's position.

Lisser declared that the kibbutz was 'of the opinion that the Arabs of Abu Ghosh and Beit Naqquba should be allowed to stay in their places . . .'. They were providing the Yishuv's forces with

[5] ISA, FM 2570/6, note on 'A telephone conversation', Shimoni to Eliahu Sasson, 15 June 1948.
[6] ISA, FM 2566/15, Y. Shimoni to B. Shitrit, 13 July 1948.
[7] ISA, FM 2564/9, 'Memorandum of Meeting on 6 July 1948 . . .'.
[8] KMA-AZP 9/9/1, 'Decisions of the Provisional Government', 30 June 1948.

information. He conceded that it was possible that they were also providing the enemy with intelligence. But he noted that, in April, the inhabitants of Abu Ghosh had voluntarily handed over to the Haganah the strategic police fort overlooking the village when the British departed.

Shitrit asked Lisser whether the Arabs of Abu Ghosh knew that the army wanted to transfer them. Lisser: 'They don't know about this—but I am preparing them for this psychologically.' At one point in the conversation, Lisser said that it was he who had proposed to Yitzhak Navon, the IDF intelligence officer responsible for liaison with Abu Ghosh and Beit Naqquba, that the villagers be transferred to Jaffa. Summarizing his position, Lisser said: 'I oppose, in principle, the transfer [of the Abu Ghosh population]—but we will not stand against the army if it decides differently.'[9]

But events were fast approaching a climax. On 6–7 July, the inhabitants of Beit Naqquba, 'including their children and women, their sheep and property (as much as they could take)' decamped, according to Shitrit 'after receiving an order from their Arab commander in Ein Karim to leave the place'.[10]

As to Abu Ghosh, '900 of the [village's] 1,000' inhabitants also left, 'though they received no expulsion order' from the Israelis (according to Shitrit). The inhabitants, wrote Shitrit, had left because of 'a state of mind' generated by the example of the evacuation of the women and children of Kiryat Anavim and the neighbouring kibbutz, Ma'aleh Hahamisha, in the days before the end of the First Truce. They were probably also influenced by the departure of their Arab neighbours from Beit Naqquba.

The army now pressed for the evacuation of the remaining inhabitants of Abu Ghosh and, on 11 July, issued a written expulsion order to this effect. Lisser, instructed by the army to tell the remaining Abu Ghosh inhabitants to leave, came to Shitrit to ask what to do. Lisser repeated his proposal that the population be transferred to Jaffa, if they could not remain in the

---

[9] ISA, FM 2564/9, 'Memorandum of a Meeting on 6 July 1948 . . .'.

[10] ISA, MAM 307/44, 'Information Concerning the Arabs of Abu Ghosh', by Bechor Shitrit, 12 July 1948, reporting on his meeting that morning with Ya'akov Lisser. The diary of Kiryat Anavim's area commander (ma'az) put the flight of the Beit Naqquba inhabitants on 8 July. See *Al Em Haderekh*, (Kiryat Anavim Press, 1949), 34.

village.[11] Shitrit thereupon wrote directly to Harel Brigade HQ saying that he had seen the order (file number 'kof-resh/7/16')— which stated that the evacuation of Abu Ghosh was 'necessary and urgent' and that 'every assistance' must be rendered to Lisser in carrying out the order. Shitrit informed Harel HQ that he had protested to Foreign Minister Moshe Sharett—obviously on the grounds that the army was acting contrary to the Cabinet decision of 30 June and without the authorization of the Ministerial Committee for Abandoned Property. Sharett, Shitrit told Harel HQ, 'informed me that the eviction should not be carried out so long as the Minister of Defence [David Ben-Gurion] had not ruled in the matter'. Shitrit ordered Harel HQ to ask for Ben-Gurion's ruling before taking any action. Should Ben-Gurion support the eviction, added Shitrit, it should be carried out with a minimum [amount of] discomfort for the inhabitants of this village . . . who, to some extent, had helped us, and we do not want it said that we demanded their evacuation'.[12]

But it appears that Lisser (or Shitrit) had not been completely frank about the reasons for the departure of the inhabitants of Beit Naqquba and of the bulk of the population of Abu Ghosh. The Harel Brigade expulsion order (of 11 July) regarding Abu Ghosh, indeed, had been formulated *after* the bulk of the inhabitants had left and, in any event, was not served. But the 900 had apparently left at least in part because they had also been strongly advised, if not actually ordered, to do so by Lisser himself and by Yitzhak Navon, the local IDF intelligence officer. Navon later recalled that he had gathered together (apparently in Kibbutz Kiryat Anavim around 5–6 July) representatives of both Abu Ghosh and Beit Naqquba and had told them that 'they must leave' (though they would be allowed back, he promised, when the fighting ended). Lisser, it appears, had told the villagers the same thing, either at the same meeting or separately.

The published version of the Kiryat Anavim area commander's diary from the period more or less confirms this, at least as regards Beit Naqquba. It states, in the entry for 8 July: 'The day before the

[11] ISA, MAM 307/44, 'Information Concerning the Arabs . . .', 12 July 1948. Shitrit passed on Lisser's explanation of why the 900 had abandoned Abu Ghosh to Shimoni at the Foreign Ministry. See ISA, FM 2564/10, Shitrit to Shimoni, 19 July 1948.

[12] ISA, MAM 307/44, B. Shitrit to Harel Brigade HQ, 12 July 1948.

end of the truce . . . Arabs from Beit Naqquba came to say goodbye. They are leaving. The parting was friendly and tearful. It is difficult to see friends in tears. But the military consideration [*hashikul hatzva'i*] is for departure . . .'. The implication is that the Yishuv's 'military consideration' was the overriding factor; the Kiryat Anavim diary contains no mention of an external Arab order for Beit Naqquba to evacuate. Nor—and I think this points to Kiryat Anavim's uneasy conscience regarding these events—does the published version of the diary contain any reference to or explanation of the evacuation of the bulk of the Abu Ghosh population.[13]

Among the Yishuv leadership in mid-July it was understood that the villagers of Abu Ghosh had been expelled by the IDF. Yosef Sprinzak, the secretary general of the Histadrut trade-union federation (later Speaker of the Knesset), heatedly condemned the 'expulsion' a few days later.[14] Mapam's Agriculture Minister, Aharon Zisling, also raised the matter in Cabinet.[15]

(Incidentally, several months later Kiryat Anavim was to deny that a 'promise'—that they would eventually be allowed to return—had been given to the villagers of Beit Naqquba: 'They left the village without receiving from us any promise regarding their future', the kibbutz later wrote the government.)[16]

In any event, Shitrit's actions (the 'halt' order, regarding the 100–150 remaining inhabitants of Abu Ghosh, to Harel Brigade HQ and his talk with Sharett in the matter) and Zisling's submission of the matter to Cabinet debate on 14 July, had had the necessary effect. Ben-Gurion specifically forbade an expulsion without his 'written authorization' and, in any case, had noted that the 'urgency' of the

---

[13] *Ha'ir*, 'Tzipor Achat Ba'etz', by Yehuda Litani, 11 Nov. 1983; interview in 1987 with Mihik Shapira, of Kibbutz Kiryat Anavim, who had heard the story of the two villages in 1948 from Lisser; and *Al Em Haderekh*, p. 34. It is possible that the inhabitants of Beit Naqquba had received *both* an order to evacuate from Arab military commanders in Ein Karim and 'strong advice' to the same effect from Lisser and Navon. But it is likely that the 'advice' given in the name of the Harel Brigade, which physically controlled the area, was the more potent of the two factors in precipitating the evacuation.

[14] See Histadrut Archive (Va'ad Hapo'el Building, Tel Aviv), protocols of the meeting of the Histadrut Executive Committee, 14 July 1948.

[15] KMA-AZP 9.9.1, 'Decisions of the Provisional Government', 14 July 1948.

[16] ISA, MAM 307/44, Kiryat Anavim to 'the Government Committee Dealing with Arabs', 30 Jan. 1949.

desired expulsion had faded as a result of the capture of the neighbouring Arab village of Suba (on 12 July). But, apparently, Ben-
Gurion had indicated that he would approve the transfer—to Jaffa,
not out of the country—of the remaining Abu Ghosh population
should a military need recur.[17]

By early August, there were some 300 villagers in Abu Ghosh—
consisting of the 100–150 who had stayed put and dozens of others
who, in dribs and drabs, had infiltrated through the Arab and Israeli
lines back to their homes. The area commander of Kiryat Anavim
noted in his diary on 19 July that '200 Arabs had returned to Abu
Ghosh'.[18] Meanwhile, the bulk of the village's 'refugees', camped
out near Ramallah, were having a bad time. The locals treated them
like traitors; a handful were even jailed.

As to the Arabs of Beit Naqquba, they had come to temporary
rest in the abandoned village of Sataf, in no-man's land, to the
south, 'under trees, because the Arabs had not allowed them to
come over their lines, out of distrust and revenge'. The IDF had
agreed to their staying at Sataf. The Beit Naqquba refugees had
appealed, via Abu Ghosh notables and Lisser, to be allowed to
return to their village or, at least, to come to Abu Ghosh. Lisser
reported that the inhabitants of Abu Ghosh were willing to take in
their brethren from Beit Naqquba.[19]

Shitrit asked Ben-Gurion for instructions.[20] The Prime Minister,
responding through Cabinet Secretary Za'ev Sharef, evaded relevant
reply: 'The people of Abu Ghosh had remained in their village',
wrote Sharef, 'with the permission of the authorities.'[21]

Meanwhile, Lisser or Kiryat Anavim were busy playing a double
game. While passing on to Shitrit the Beit Naqquba refugees' request
to return, Lisser or fellow members from Kiryat Anavim had
informed the army that the Beit Naqquba villagers were infiltrating
into Abu Ghosh and settling in its abandoned houses. Kiryat

[17] ISA, FM 2564/10, Shitrit to Shimoni, 19 July 1948.
[18] *Al Em Haderekh*, p. 35.
[19] ISA, FM 2564/9, memorandum by Shitrit after a conversation with Lisser, 9
Aug. 1948.
[20] ISA, FM 2564/9, the Minister for Minority Affairs to the Prime Minister of
Defence, 16 Aug. 1948.
[21] ISA, FM 2564/9, Cabinet Secretary to Minister for Minority Affairs, 30 Aug.
1948; ISA MAM 307/44, unsigned minute 'to the minister', 30 Aug. 1948; and ISA
MAM 307/44, E. Chelouche to the Cabinet Secretariat, 9 Sept. 1948.

Anavim apparently hoped that the army would act to halt the influx of these refugees. Moreover, the kibbutz 'also proposed that the remaining Abu Ghosh villagers be transferred to Jaffa and thus end the problem of Arabs [living] on the main [Jerusalem–Tel Aviv] road, but this proposal did not at the time receive the requisite attention. The Arabs of Abu Ghosh, without hindrance, began to return to their village and with them, the Arabs of the village of Beit Naqquba.'[22]

Shitrit apparently got no direct reply from Ben-Gurion concerning the refugees' request to return. But the army, in the person of General Zvi Ayalon, OC Central Front, ruled in early September that it 'would not allow additional Arabs to enter Abu Ghosh as every additional Arab in our area is a superfluous burden'.[23] This view was shared by General Allon, the commander of the Palmah, and by Tabenkin, the commander of the Harel Brigade. Their attitude was that the Jerusalem corridor should remain 'clear of Arabs', at least for many years to come.[24]

Ayalon's attitude had in part been prompted by the continuous 'illegal' trickle during July and August of the Abu Ghosh refugees back to their home. This trickle also raised the hackles of Shimoni at the Foreign Ministry, who wrote to Shitrit that while he, too, was awaiting the Prime Minister's ruling regarding the refugees from both villages, the matter must not be solved by the unilateral, piecemeal return of the villagers as infiltrees. This 'disorganized' return must be halted.[25] But Lisser and Navon continued to lobby in Central Front HQ for formally allowing a return of the refugees, at least to Abu Ghosh (as the two had promised the departees in July).[26]

Meanwhile, many of the refugees from Abu Ghosh—the local army commanders apparently turning a blind eye—slowly made their way back to the village. By mid-January 1949, it was reported that the village, which had had 960 inhabitants before the war, now

[22] ISA, MAM 307/44, Kvutzat Kiryat Anavim to 'the Government Committee Dealing with Arabs', 30 Jan. 1949.
[23] ISA, MAM 307/44, HQ Military Government in the Occupied Areas to Minister for Minority Affairs, Defence Minister, Foreign Minister, 5 Sept. 1948.
[24] Interview in 1987 with Mihik Shapira, of Kibbutz Kiryat Anavim, who had heard from Lisser the Abu Ghosh and Beit Naqquba story.
[25] ISA, FM 2564/9, Y. Shimoni to the Minister for Minority Affairs, 25 Aug. 1948.
[26] Interview in 1987 with Mihik Shapira.

had 600, which included 60 originally from Beit Naqquba and 50 from another smaller, abandoned satellite village, 'Ein Rafa.[27]

By early 1949, Kiryat Anavim appears to have resigned itself to the permanent repopulation of its neighbour, Abu Ghosh. But it persistently lobbied against the government allowing the Beit Naqquba refugees to be allowed to stay in Abu Ghosh or to be allowed to return to their original village: 'In our opinion, it is illogical to allow the return of the inhabitants of the village [of Beit Naqquba, resident in Abu Ghosh, 97 in number] for the following reasons: (*a*) The village of Beit Naqquba is indeed too close to our kibbutz. (*b*) If the village is "renewed", our [kibbutz] will be surrounded on two sides.'

Kiryat Anavim proposed the resettlement of the refugees of both Beit Naqquba and Abu Ghosh in the nearby abandoned village of Khirbet al 'Umur. (It was unclear whether the kibbutz, when speaking of Abu Ghosh 'refugees', was referring only to those villagers who were still living outside the village or also to those who had remained throughout or had already returned to Abu Ghosh.)[28]

Kiryat Anavim's opposition to the return of the Beit Naqquba refugees to their village was strongly backed by the local Jewish regional council, on which sat representatives of Kibbutz Ma'aleh Ha'hamisha and other settlements.[29]

Kiryat Anavim's stand was only in part based on 'security' considerations. The kibbutz also wanted Beit Naqquba's lands. The problem was that the handful of Beit Naqquba refugees now living in Abu Ghosh continued to cultivate their lands, 'and it is to be assumed that they look forward to the day on which they will be able to return to their homes. It seems that so long as the Beit Naqquba inhabitants remain near their abandoned village, they will continue to maintain contact with their village, and the members of Kiryat Anavim will not be able to take over and cultivate the village lands.' Reporting this, the Interior Ministry official responsible for the Jerusalem District recommended that the Beit Naqquba villagers

[27] ISA, MAM 307/44, Shlomo Asherov to Gad Machnes, 17 Jan. 1949.
[28] ISA, MAM 307/44, Kvutzat Kiryat Anavim to 'the Government Committee Dealing with Arabs', 30 Jan. 1949.
[29] ISA, MAM 307/44, Judean Hills Settlements Bloc Committee to the Minister for Minority Affairs, 12 Nov. 1948.

residing in Abu Ghosh be moved 'somewhere . . . far away'.[30]

There remained a further problem—the nearby satellite village of 'Ein Rafa. Some ten families—'35 souls'—had either never abandoned it or had returned to their homes during the preceding months. The Interior Ministry official noted that the place had become a way-station for Arab infiltrees through the Jerusalem corridor and that 'the men of the new [Jewish] settlement of Misgav Palmah [subsequently named Kibbutz Tzuba, next to the abandoned Arab site, Suba] were very interested in ['Ein Rafa's] lands'. The official recommended that the ten families be moved to 'another Arab area'.[31] But his recommendation was never acted upon. 'Ein Rafa remains to this day.

Over the months, the bulk of Abu Ghosh's inhabitants returned to their homes, the army and civil authorities in this case turning a blind eye to the influx of Arabs. But sometime during the second half of 1949, as part of the government's increasingly hard-line policy *vis-à-vis* Arab infiltrators, the authorities decided to put an end to the continued, piecemeal infiltration of Abu Ghosh and Beit Naqquba refugees back to Abu Ghosh. (All such illegally returning refugees were regarded as illegal residents, and liable to arbitrary deportation.) Those who had made it back to Abu Ghosh by March–April 1949 were left alone (even though they had not been in the country during the autumn 1948 census and were not necessarily in possession of official ID cards). But the more recent infiltrators were targeted for expulsion. In a series of more or less brutal search-and-expulsion operations, police and IDF units descended upon the village, identified and rounded up infiltrees, and pushed them over the border into Jordan.

Following one such round-up, in early July 1950, the inhabitants of Abu Ghosh sent off an emotional appeal to Knesset members, political parties, and journalists. In their 'open letter', the inhabitants charged that the Israelis had repeatedly 'surrounded our village, and taken our women, children and old folk, and thrown them over

---

[30] ISA, MAM 307/44, A. Bergman (Biran), Jerusalem District Commissioner, to 'the Government Committee for the Settlement of Dispossessed Arabs', 16 Mar. 1949.
[31] ISA, MAM 307/44, A. Bergman, Jerusalem District Commissioner, to 'the Government Committee for Settling Dispossessed Arabs', 16 Mar. 1949.

the border, and into the Negev Desert, and many of them died in consequence, when they were shot [trying to make their way back across] the borders'. So far, the inhabitants had held their peace.

But we cannot remain silent in face of the latest incident last Friday, when we woke up to the shouts blaring over the loudspeaker announcing that the village was surrounded and anyone trying to get out would be shot . . . The police and military forces then began to enter the houses and conduct meticulous searches, but no contraband was found. In the end, using force and blows, they gathered up our women and old folk and children, the sick and the blind and pregnant women. These shouted for help but there was no saviour. And we looked on and were powerless to do anything save beg for mercy. Alas, all our pleas were of no avail . . . They then took the prisoners, who were weeping and screaming, to an unknown place, and we still do not know what befell them.

On 11 July, taking heed of the press and public criticism that the affair had sparked, which included a Knesset debate and vote (won by the coalition), Ben-Gurion jotted down in his diary:

[Jerusalem District Commissioner Avraham] Biran and the chief of police came to me concerning the round-up in Abu Ghosh. In June [should be, July] 1948 there were 180 souls in the village . . . Today there are more than 800 souls. In the round-up, 105 persons were caught, mostly children and women, [and] a minority of men. That is the method of infiltration—they first send women and children, afterwards comes the male. There are thieves and smugglers in the village. According to [Moshe] Perlman, there are 300 infiltrators in the village; 150 were left alone, 150 must be expelled; only 105 were caught, there are another 45 [to catch and expel] . . . There were no acts of cruelty in the round-up. The infiltrators caught were handed over to the army for expulsion.

The incident also sparked some soul-searching within the Cabinet. Foreign Minister Sharett wrote that

with all the logical and relevant justification for the wholesale round-up, which included women and children, it is a fact that forcible expulsion of women and children stirs up the public and perturbs its conscience. The public will not countenance such measures for long and it is well to take note of this . . . Therefore, the battle against the infiltrators must be waged in ways that will not lead to a public rebellion against the government . . .

Sharett proposed that the government act with greater vigour and immediacy against infiltrators, so that they do not amass in each village and do not manage to reside in the village for a year and longer before the authorities act against them. He also cautioned against expelling women and children; 'only adult males' should be expelled.

Sharett was able to discover a silver lining in the affair: 'One must comment that the fact that there was, among the public, a display of moral sensitivity is praiseworthy and I am certain that none of us would like to see in our public a complete lack of moral sensibility regarding such issues of humanity.'[32] In the end only several dozen Abu Ghosh families remained in exile, as refugees, in the Ramallah area in the West Bank.

But as the months wore on, the Beit Naqquba 'refugees' living in Abu Ghosh outwore their welcome. The Abu Ghosh leaders wanted them out. But their homes in Beit Naqquba 'had been given to new [Jewish] immigrants from Yugoslavia' and their orchards—apples, pears—were being cultivated by Jews 'who were not allowing the Arabs of Beit Naqquba to enjoy the fruits of their trees'. The Agriculture Ministry recommended that they be resettled near Lydda or Ramle, where there were empty, abandoned, semi-intact Arab houses. But the Arabs of Beit Naqquba refused to budge, demanding that they be resettled on their lands some distance from Abu Ghosh and from their original village, to which, they gradually understood, they could never return. So, by common consent, the Beit Naqquba refugees were allowed to stay on in Abu Ghosh.[33]

For fourteen years the refugees from Beit Naqquba lived in Abu Ghosh. Then, starting in 1962, with the permission and encouragement of the Israeli authorities, they gradually moved to a new site, named 'Ein Naqquba, on some of their lands south of the Jerusalem–Tel Aviv highway. After 1967, some Beit Naqquba refugees who had lived in refugee camps and villages in the West Bank, around Bethlehem, were also allowed back into Israel, and joined their

[32] 'An Open Letter to the Inhabitants of Israel', 25 Ramadan 1369 (10 July 1950), by the inhabitants of the village of Abu Ghosh, in HHA kaf-90-1/6; Ben-Gurion diary, 11 July 1950, in David Ben-Gurion Archive, Sdeh Boqer; and Foreign Minister Moshe Sharett to the advisor on special affairs, Reuven Shiloah, 14 July 1950, in ISA, FM 2402/12.

[33] ISA, AM 591, Avraham Hanuki, secretary of the Agriculture Ministry, to 'the Committee for Refugees next to the Prime Minister's Office', 2 Nov. 1949; A. Hanuki to the Adviser on Arab Affairs, the Prime Minister's Office, 28 Feb. 1950; and Moshe Levin, in the name of Zalman Liff (Lifshitz), the Adviser to the Prime Minister on Land Matters, to the Custodian of Absentees Property, 16 Mar. 1950.

brethren at the new village site. The site overlooks their old village site, today the Jewish settlement of Beit Nekofa.[34]

AL FUREIDIS AND KHIRBET JISR AZ ZARKA

In May 1948, the large Arab village of Al Fureidis south of Haifa and the small, former beduin village of Khirbet Jisr az Zarka ('Arab al Ghawarina), five kilometres to the south, along the Tel Aviv–Haifa highway, surrendered to the Haganah without a fight. The surrender had been preceded by orders to Al Fureidis, in late April, by the Arab Liberation Army, the foreign Arab volunteer force assisting the Palestinians against the Yishuv, to evacuate its 'women and children . . . and to be ready to completely evacuate [the village]'.[35] The local Arab command was clearly worried by the prospect that the village would be attacked by the Haganah and fall, and/or that the inhabitants might agree to surrender and live under Jewish rule.

When the Arab inhabitants of the northern Coastal Plain began to leave their villages in April, out of panic, or under Jewish or Arab pressure, the inhabitants of Al Fureidis and Khirbet Jisr az Zarka 'also wanted to abandon their villages out of fear of coming to harm, but under the influence of the heads of the *moshavot* [Jewish private farmers' settlements] of Zikhron Ya'akov and Binyamina, they stayed in their places, after being promised that they had nothing to fear'.[36]

Before 1948, relations between Al Fureidis and Khirbet Jisr az Zarka and the neighbouring Jewish *moshava* of Zikhron Ya'akov, one of the country's main wine-producing centres, had been generally good. The villages supplied the *moshava* with field-hands for its vineyards. During the first months of the war, snipers from Al Fureidis occasionally fired on Jewish traffic on the main road. But in general, peace prevailed between Zikhron and its Arab neighbours.

By early May, with the Haganah clearly on the ascendant, Al Fureidis (and two nearby villages, As Sindiyana and Sabbarin)

[34] *Ha'ir* 'Tzipor Achat Ba'etz', by Yehuda Litani, 11 Nov. 1983.

[35] KMA-PA 100/MemVavDalet/3—150, 'Yediot Tene, 6.5.1948.', referring to 24 Apr. 1948.

[36] ISA, MAM 307/31, the Minister for Minority Affairs to the Ministers of Defence, Agriculture, and Labour, 25 June 1948. Shitrit was reporting on a meeting he had had on 24 June 1948 with local leaders from Zikhron Ya'akov and Binyamina.

decided, according to Haganah intelligence, 'not to allow the entry of foreign [Arab] forces and not to serve on any account as a base for the [Arab] Liberation Army . . .'. Moreover, Al Fureidis was 'willing to live in peace with its [Jewish] neighbours . . .'.[37] Within days, the village surrendered to the Haganah, giving up its arms.

The inhabitants of Khirbet Jisr az Zarka had surrendered to the Haganah a few days earlier. A meeting of Haganah intelligence officers immediately afterwards ruled against expulsion, instructing that 'the village is to be treated in accordance with the guidelines laid down regarding a surrendering village'.[38]

But within weeks, the presence of the two villages in the northern Coastal Plain, in an area completely cleared of Arabs during the previous months, began to irk the local IDF unit, the Alexandroni Brigade. The IDF, complained Abba Shechter, the 'mayor' of Zikhron, Shalom Burstein, a member of the vine-growers association, and Asher Schumann, the chairman of the Agricultural Committee in Binyamina, wanted to prohibit the inhabitants of the two villages from 'moving around'. The problem, the delegation of local Jewish leaders told Shitrit, was that Al Fureidis (with a population of some 500–600) supplied the two *moshavot* with 129 workers; Khirbet Jisr az Zarka supplied an additional, smaller number. 'The agricultural work in Zikhron Ya'akov and Binyamina is based mainly on hired [Arab] labour.' In the past, workers from the two villages had worked in the Zikhron and Binyamina fields together with youngsters from the *moshavot*. But the Jewish youngsters were all mobilized in the army, 'and now the farming in the two *moshavot* is dependent on the work of the Arab labourers from the nearby [villages]. If these workers are expelled or arrested, the farms will remain without workers and agriculture will be seriously jeopardised.'

The Alexandroni Brigade OC, Dan Epstein, and the district commander, Yitzhak Shemi, were now preventing the villagers from reaching the Jewish fields and vineyards. The local Jewish leaders had proposed to Shemi that the villagers be restricted to their villages by curfew during the nights but that they be allowed to

[37] KMA-PA 100/MemVavDalet/3—154, 'Yediot Tene 9 May 1948'; see also ISA, MAM 307/31, Minister for Minority Affairs to the Ministers of Defence, Agriculture, and Labour, 25 June 1948.

[38] David Ben-Gurion Archive, 'Summary of the Meeting of Arab Affairs Advisers in Netanya', 9 May 1948.

go to work in the daylight hours, after being issued with ID cards. 'But he naturally rejected this', the Jewish leaders complained. And, above everything, hovered the threat of a permanent solution —a mass expulsion. If the villagers are expelled, 'the heads of the *moshavot* will appear as breakers of promises'. Moreover, the promises were given, said the local Jewish leaders, only after they 'had consulted with the defence forces'. In other words, the local IDF commanders had approved the Zikhron and Binyamina leaders' appeals to the inhabitants of Al Fureidis and Khirbet Jisr az Zarka to stay and, at least implicitly, had endorsed the assurances that no harm would come to them if they did so.

The local leaders were apparently persuasive. Two days after their meeting with the Minister for Minority Affairs, Shitrit appealed to his fellow ministers, including David Ben-Gurion, 'to delay for four weeks any action aimed at the removal of the inhabitants of these villages or their detention as prisoners [an option apparently threatened by the Alexandroni Brigade officers] . . . .'.[39]

Shitrit won the support of Mapam. Mordechai Bentov, the Minister of Labour, proposed that the police (who were controlled by Shitrit himself) be placed in charge of the two villages. 'I don't think the security forces have any legal basis for carrying out these mooted actions [that is, expulsion or detentions]', he said.[40]

The following day the matter came up for decision in the Cabinet, apparently tabled by Shitrit. There was 'a plan to transfer the Arabs of Fureidis and Ghawarina'. The ministers decided that Arabs were not to be 'transferred from place to place without the decision of the committee for the abandoned area', that is, the Ministerial Committee for Abandoned Property, which met for the first time a fortnight later.[41]

But the approaching end of the First Truce—as with the Harel Brigade and Abu Ghosh and Beit Naqquba—fired the Alexandroni Brigade into action. On 7 July, the day before the truce ended, Epstein decided 'to end the Ghawarina affair. He intends to transfer them to the Arab area [that is, outside the state] or to a prisoner-of-war camp.' The reason for the precipitate decision, according to the

[39] ISA, MAM 307/31, the Minister of Minority Affairs to the Ministers of Defence, Agriculture, and Labour, 25 June 1948.
[40] ISA, MAM 307/31, M. Bentov to B. Shitrit, 29 June 1948.
[41] KMA-AZP 9.9.1., 'Decisions of the Provisional Government', 30 June 1948.

brigade's Arab-affairs adviser, Shlomo Zucker, was that 'a gang had visited and run amok' in Khirbet Jisr az Zarka. The adviser despatched an 'urgent' cable to the 'IDF General Staff / the Department for Arab Affairs' to step in and halt the prospective expulsion.[42]

Shitrit, to whom the adviser had sent a copy of his cable, the same day contacted Alexandroni HQ to try to find out what was going on, but failed to locate the relevant officers. His only comment was that he opposed transferring the inhabitants of Khirbet Jisr az Zarka to Al Fureidis—an option not actually mentioned in the adviser's cable.[43] Two days later the adviser clarified matters: 'The gang's rampage . . . was directed against villagers and was aimed at robbery.' (The gangsters got away with I£700). In his confused response, the adviser seemed to express support for transferring the Ghawarina, should the need arise. But, meanwhile, expulsion had been stymied.[44]

No sooner had the problem of the Ghawarina died down, however, than the bigger problem of Al Fureidis surged again to the surface. In mid-July, the IDF was heavily engaged on all three fronts against the Arab armies—the Arab Liberation Army and the Syrian Army in the north, the Arab Legion in the centre, and the Egyptian Army in the south. The resulting tension also encompassed Alexandroni, which remained fairly inactive but poised to repel a possible thrust towards the coast by the Legion or the Iraqi Army stationed in Samaria. Alexandroni clearly wanted Al Fureidis—behind the lines and sitting astride a major highway—out.

In early July, the brigade's Arab-affairs adviser journeyed to Jaffa—where the IDF Intelligence Service was headquartered—to take advice. He then recommended, in view of the lively and 'undesirable' illegal movement of infiltrators back and forth around Al Fureidis, that ID cards be issued to the inhabitants, that a police

[42] ISA, MAM 307/48, Arab-affairs adviser / Alexandroni Brigade, to General Staff / Department for Arab Affairs, 8 July 1948.

[43] ISA, MAM 307/48, Shitrit to the adviser for Arab affairs, Alexandroni Brigade, 9 July 1948.

[44] ISA, MAM 307/48, Arab-affairs Adviser / Alexandroni Brigade to the Minister for Minority Affairs, 12 July 1948. Mapam may also have had a hand in removing the expulsion threat. At the meeting of the Histadrut Executive Committee on 14 July, Moshe Erem, a Mapam representative, referred to 'an Arab village near Zikhron Ya'akov' that, 'with great trouble, we succeeded' in saving from expulsion, 'for the time being'. See the protocols of the Histadrut Executive Committee meeting of 14 July 1948, Histadrut Archive.

station be established in the village, and that a permanent nightly curfew be imposed.[45] This advice was promptly acted upon: Alexandroni notified the Agricultural Committee at Zikhron, Binyamina, and Hadera that it intended to institute a curfew in Fureidis and Jisr az Zarka 'and to prevent them by this means from going to work in Zikhron and its environs unless the farmers pay the army 580 [should read 850] mil per day as wages [1,000 mil = I£1], from which 500 mil will be paid the Arab worker and 350 mil will go to the army'.[46] The meaning was clear: either the Jewish farmers pay the army 350 mil per day per Arab worker or the army would completely seal off the villages, thus depriving the *moshavot* of their vital field hands.

The Zikhron farmers appealed once again to the Ministry for Minority Affairs. The implementation of the edict 'would do grave harm to the agricultural work of the *moshava*'. The farmers demanded that the edict be annulled or that its implementation be deferred 'at least until after the grape-gathering'.[47]

Shitrit raised the matter at the meeting of the Ministerial Committee for Abandoned Property the same day. The ministers ruled that 'this matter is under the jurisdiction of the Minister for Minority Affairs. The committee rejects [the idea] . . . and empowers the Minister for Minority Affairs to act accordingly.'[48] Shitrit then wrote to Alexandroni, expressing the ministerial committee's

astonishment . . . If the presence of the Arab workers endangers security, then the 350 mil the army will receive for each worker will not eliminate this danger, and if their presence is not dangerous to security, then there is no reason—indeed, it is prohibited—to impose on the [Jewish] farmers a payment, which is in the nature of a tax . . . Such an action is certainly illegal.

[45] ISA, MAM 307/48, Arab-affairs adviser / Alexandroni Brigade to IDF General Staff, etc., 10 July 1948.

[46] ISA, MAM 307/31, Minister for Minority Affairs to Alexandroni / Shlomo Zucker, 13 July 1948.

[47] ISA, MAM 307/48, H. Ariav, Secretary General of the Farmers Association of the Land of Israel, to Gad Machnes, the Director General of the Ministry for Minority Affairs, 13 July 1948.

[48] ISA, FM 2401/21 aleph, 'Protocol of the Meeting of the Ministerial Committee for Abandoned Property that Took Place in the Kiriya [the Government compound, Tel Aviv], on Friday, 13 July 1948'.

Shitrit went on to forbid the brigade from 'uprooting any Arab worker from his place without an order to this effect from the Minister of Defence'.[49]

To recap for a moment: in late June, the Alexandroni Brigade had sought to expel the inhabitants of Al Fureidis and Khirbet Jisr az Zarka. But prompt intervention by the agricultural notables of Zikhron and Binyamina, driven by economic considerations, and the support of the Minority Affairs Minister—who obtained Cabinet backing—had blocked an expulsion. Yet Alexandroni still wanted the two thorns in its side removed. No doubt the constant movement of Arabs to and from the two villages—in part between the villages and areas to the north still under Arab occupation, such as the Al Jaba–Ijzim–'Ein Ghazal triangle—perturbed the brigade. The outbreak of fresh fighting on 8–9 July, with the end of the First Truce, seemed to provide an opportunity. The brigade informed the Jewish farmers that the presence of the two Arab villages and the free movement of its menfolk to and from work in the Jewish fields and vineyards was troublesome. The villages would either have to be placed under a round-the-clock curfew or the Jewish farmers would have to pay the army for its security services. The thinking behind this seems to have been that in either case, the Zikhron and Binyamina farmers would be forced to adjust to the idea of dispensing with the Arabs' labour: either the curfew would physically keep the field-hands away or the 'tax' would make their services prohibitively expensive. The local Jewish farmers could then be expected to lose interest in the fate of Al Fureidis and Khirbet Jisr az Zarka and Alexandroni would be free, at last, to expel the villagers.

Shitrit's letter of 13 July failed to impress Alexandroni's commanders. Two days later they placed Al Fureidis and Khirbet Jisr az Zarka under quarantine, barring all entry to and exit from the two villages until further notice. The brigade's Arab-affairs adviser promptly complained to Shitrit that a local commander had acted without his knowledge or approval and registered his disagreement with the action: the Ghawarina and Fureidis

are not so dangerous that they must be placed under complete quarantine . . . It is possible that there was need for some sort of action to increase

[49] ISA, MAM 307/31, Minister for Minority Affairs to Alexandroni / Shlomo Zucker, 13 July 1948.

their isolation from the unconquered villages [Jaba, 'Ein Ghazal, and Ijzim] and the Triangle [the Palestinian-populated area of Nablus, Jenin, and Tulkarm] but this could have been done in a more fair and humane way.[50]

Meanwhile, the deputy brigade commander, Binyamin Goldstein, also wrote to Shitrit (and to Ben-Gurion), explaining Alexandroni's actions. The two villages, he declared,

endanger security in the district, through constant infiltration to and from them. Security [considerations] compel me to impose constant surveillance on them, and this can be done in two ways: (A) To remove them from our area to the Arab-held area. (B) To keep them under constant surveillance . . . I had to take the second course . . . The difference in the wages will cover the costs of supervising them.

Alternatively, wrote Goldstein, the villagers could be 'held as prisoners-of-war', and could be employed by the Office of the Custodian for Absentees Property picking fruit or reaping crops. The problem, implied Goldstein, was as much economic as military: without work, the villagers would be penniless and would have to be maintained by the army. Alexandroni lacked the funds for this.[51]

Shitrit backed down. Perhaps Ben-Gurion had told him to lay off Alexandroni, it was military business. But the quarantine galvanized the Zikhron farmers into action. A delegation of plaintiffs duly arrived at Shitrit's doorstep on 16 July.

Shitrit. I raised the matter in the Cabinet and sent [Alexandroni] a letter.

The farmers' delegation. From yesterday they imposed a quarantine on the village. What does the minister think?

Shitrit proposed that they pay out the 850 mils per day per worker (with the army getting 350 mils) and that the farmers then ask the Treasury and Shitrit himself for a refund. 'I'll fix the matter', he promised. 'If the Treasury won't pay you, I tell you to go ahead and sue [the government or army].'

The farmers were not pleased. They wanted instant action: 'The vineyards are ready for harvesting . . .'.

---

[50] ISA, MAM 307/48, Arab-affairs adviser / Alexandroni Brigade to Minister for Minority Affairs, 15 or 16 July 1948.

[51] ISA, MAM 307/48, Binyamin Goldstein, Alexandroni Brigade, to the Minister for Minority Affairs, the Prime Minister and Minister of Defence, and the Minister of Finance, 17 July 1948. Zucker, in his letter to Shitrit of 15 or 16 July, had written that Fureidis 'lacked provisions to maintain themselves even for one week'.

Shitrit. I'll write today to the Prime Minister.[52]

The Zikhron farmers must have felt that not much had been achieved. Niether was the Minority Affairs Ministry's effort to mobilize the Finance Minister, Eliezer Kaplan, against Alexandroni any more successful. 'The question . . . was raised in Cabinet . . . I will be happy to learn from you what is the situation', Kaplan wrote vaguely and irrelevantly a few days later to Machnes, the Director General of the Ministry for Minority Affairs.[53]

The affair, at this stage, was also to have repercussions within Mapam. It was raised by both Meir Ya'ari and Moshe Erem, party leaders, at a meeting on 26–7 July of Mapam's defence officials and senior IDF officers. Erem said that the quarantine of the two villages meant that they would lack for food. He blamed an unnamed party member—who was a senior officer in Alexandroni, perhaps Goldstein—for what was happening. The unnamed Mapam officer had told him that the villages constituted 'a Fifth Column' and that supplying them with food was like 'giving supplies to the enemy'. Yet Zucker, said Erem, 'who was not a member of our party', thought that the whole quarantine was 'exaggerated and that there is no need to seal them off'.[54]

A fresh and ominous twist to the affair was introduced on 26 July, when two General Staff / Operations officers appeared at the district headquarters in Netanya and told the Alexandroni staff to 'prepare to transfer the inhabitants of [Fureidis and Khirbet Jisr az Zarka] to the Arab[-held] area at 1800 hours'. The local district commander, 'Naftali' (perhaps Shemi), 'refused and said that he must receive the order from his direct superior Binyamin [Goldstein]'. Goldstein, for his part, delayed carrying out the order until consulting with Zucker, the Arab-affairs adviser. Zucker 'opposed the operation' and promptly informed Shitrit of what was going on.[55] The expulsion was stymied.

[52] ISA, MAM 307/48, 'The Minister's Interview with a Delegation from Zikhron Ya'akov', 16 July 1948.

[53] ISA, MAM 307/48, Gad Machnes to the Minister of Finance, 18 July 1948; and E. Kaplan to Machnes, 26 July 1948.

[54] Hashomer Hatzair Archive 10.18, 'Protocols of the Meeting of Mapam Defence Activists', 26–7 July 1948.

[55] ISA, MAM 307/48, Arab-affairs adviser / Alexandroni Brigade to the Ministry for Minority Affairs, 27 July 1948.

But the IDF quarantine and its possible alternative, the 850 mils arrangement, remained. At the end of July, following the end, on 18 July, of the second round of Arab–Israeli hostilities and a change in the strategic situation in the area, a compromise agreement was at last hammered out between the army and the local Jewish farmers, with the Minority Affairs Ministry advising, and perhaps mediating, from the sidelines. The subject of the agreement—the inhabitants of Al Fureidis and Khirbet Jisr az Zarka—did not participate in the discussions. The crucial meeting took place on 27 July, with the chairmen of the Zikhron Ya'akov and Binyamina local councils (Abba Shechter and M. Paicovitch), the IDF district commander (Yitzhak Shemi), the district police commander (Yitzhak Ressner), the IDF district intelligence officer (Binyamin Winter), and local kibbutz leaders participating.

All present noted the history of good relations between the two villages and their Jewish neighbours and the villages' resistance to the Arab 'gangs'. The main security problem during the previous weeks had been the infiltration back and forth between the two villages and the still unconquered Arab villages of the so-called 'Little Triangle' (Jaba, 'Ein Ghazal, and Ijzim). But the 'Little Triangle' had fallen to Jewish attack earlier that week; the villagers had fled and / or been driven out. The problem of infiltration between the hostile villages of the 'Little Triangle' and Al Fureidis and the Ghawarina had thus disappeared. The situation had changed.

The local leaders and the IDF district commander, Shemi, agreed that henceforward, the inhabitants of the two villages would be allowed to work in Zikhron Ya'akov and Binyamina in the vineyards and the wine press; they would be transported to and from work in organized fashion in vehicles; the adult male hands would be paid 600 mils per day (with '400 mils for women, [and] 300 mils for children'); the Jewish farmers would be responsible for and pay the costs of the transportation; the farmers would pay the army an additional 75 mils per male, adult hand 'for guarding the village[s], and for providing supplies for and development of the village[s]'. The money was to be handed to IDF-appointed supervisors whose work was to be regulated by the Minority Affairs Ministry. The two villages were to receive their supplies from Zikhron Ya'akov and Binyamina until shops were set up in the villages. The *moshavot* were also responsible for the medical care of the villagers. The villagers were to be allowed to cultivate their fields (from which the quarantine had previously barred them).[56]

At the beginning of August, Erem visited the area to see how the agreement was being implemented. 'The workers are travelling to work every day in organised fashion, the supervisors registering the workers when they leave [home] and upon their return.' The supervisors were paid the 75 mils per head per day. During his visit, Erem encountered a truck 'bringing the workers back to Al Ghawarina; it was full and crowded [with] (70 persons), to the point of danger . . .'. Erem reprimanded the driver; the Arab passengers, for their part, complained that men and women were being transported tightly packed together—'they feared possible harm to family relations'. In Fureidis, Erem learned that beside the village's 700 permanent inhabitants, the village contained 215 refugees from nearby Tantura, conquered by the Haganah and abandoned by its inhabitants in May. The refugees were receiving their supplies from the Fureidis villagers and at their expense. The refugees had no housing and lived out in the open. The village had no school or teachers. Twice a week, it was visited by a Jewish doctor. The villagers complained that they had sown wheat in the abandoned fields of Tantura but that the army had then reaped the crop.

In Binyamina, the Jewish farmers complained that the villagers did not need the 'busing' and could make their way to the fields by foot. 'They argued that the security situation west of the *moshavot* was not problematic and therefore the matter of transport was superfluous.' They also thought the 75 mils charge unnecessary. But Erem said the transportation arrangement and the 75 mils payment were irreversible. He summed the matter up by saying that 'the arrangement added a pleasant chapter to the [history of] good neighbourly relations [in the area]'. Binyamina and Zikhron Ya'akov were thus providing funds for the development of the two Arab villages. The meeting between Erem and the Binyamina farmers

---

[56] ISA, MAM 307/48, 'Memorandum on the Meeting with the Shomron [Samaria] Settlements Bloc Concerning the Inhabitants of the Villages of Fureidis and Al Ghawarina', by Moshe Erem (the Mapam leader and Minority Affairs Minister official responsible for the agreement), 1 Aug. 1948 (the meeting took place on 27 July 1948).

ended on a cheerful note, Erem recorded: Binyamina would carry out the agreement.[57]

Thus was resolved the 'problem' of Al Fureidis and Khirbet Jisr az Zarka. The army had failed to dislodge the two villages, which to this day, having grown and prospered, overlook the Tel Aviv–Haifa highway.

CONCLUSION

The stories of Abu Ghosh and Beit Naqquba and Al Fureidis and Khirbet Jisr az Zarka in 1948 are intructive in a number of ways about Jewish–Arab relations and about Yishuv decision-making concerning Arab communities during the first Israeli–Arab war.

One notes clear, common patterns. In both areas, the local Haganah / IDF command (Harel and Alexandroni brigades), in varying degrees backed by the high command, strongly desired the removal of the villages, which were regarded by the military as thorns in their side. They wanted Arab-clear areas behind the front lines and along the vital roads under their supervision. In both areas, during the First Truce and, even more pressingly, during the subsequent 'Ten Days' of hostilities (9–18 July), the army made efforts, albeit falling short of outright expulsion, which was forbidden by the political echelon, to obtain the removal of the villages.

In the Jerusalem corridor, the tactic was one of pressure and 'advice' to leave, with the military using the local 'good guys'— Lisser and Navon—to persuade the villagers to decamp. The inhabitants of Abu Ghosh, and perhaps also those of Beit Naqquba, were promised that they would be allowed to return to their homes when the hostilities subsided. The villagers of Beit Naqquba were apparently—simultaneously—ordered by local Arab commanders to evacuate. Subsequently, the army tried, by and large unsuccessfully, to bar the return of the villagers—from both sites—to their homes in Israeli territory.

In the coastal plain, the local command and high command alternated between efforts to expel the inhabitants of Al Fureidis and Khirbet Jisr az Zarka outright and more subtle methods of

[57] ISA, MAM 307/48, memorandum on 'The Visit to Fureidis and the Ratification of the Agreement by the Binyamina Agricultural Committee', by Moshe Erem, 2 Aug. 1948.

making the villagers' continued life under Israeli rule unpleasant or impossible. (Not being allowed to work in the neighbouring Jewish fields or to cultivate their own fields would have meant economic strangulation.) The quarantine and its alternative, the 350 mil tax, both seem to have been ultimately designed to obtain the villagers' removal. In Al Fureidis and Khirbet Jisr az Zarka—far less of a war zone in July than the Jerusalem corridor—the army proved completely unsuccessful.

In both areas, the Arabs' *summud* (to use a later term for holding tight to their lands)—when backed by important local or national elements within the Yishuv—proved stronger than the army's will to uproot them.

In general, in 1948, where Arab communities fled at the first hint or whiff of grapeshot, there was no problem and the Yishuv was happy. But where Arab communities held tight *after the end of hostilities in a specific area*, the Jewish establishment, which had no master plan or agreed and public policy, and was riven by dissent, occasionally proved unavailing. There were a handful of politicians and officers who as a matter of course on political and / or moral grounds opposed expulsion. There were many others who, though they might have been satisfied, not to say happy, to see Arabs flee in panic, would have been embarrassed to be seen to be involved in forcibly uprooting Arab communities.

Happily for the Yishuv, most of the Arabs who became refugees in 1948 fled out of fear or panic before or during Haganah / IDF attack on their villages, thus freeing the Jewish officers and officials at each site from confronting the dilemma of expelling or leaving in place an Arab population. One can fairly say that when Arab communities in 1948 did not flee and / or were not expelled before or during their conquest, and the issue came up for debate in civilian bodies, meaning the Cabinet or its committees, and in the Jewish political parties, then ideology, morality, and embarrassment coalesced to stymie the army's desire to expel (where that existed). Outright expulsions such as occurred along the Lebanese border in November (Iqrit, Bir'im, Nabi Rubin, Tarbikha) were carried out by the army *without Cabinet knowledge, debate, or approval* —though, almost inevitably, they received *post facto* Cabinet endorsement. ('Eilaboun, from which the inhabitants were expelled by Golani Brigade units at the end of October, was an exception, due largely to the villagers' Maronite faith and affiliations: there, subsequently, the Cabinet approved the exiled villagers' return).

In the cases dealt with here, clearly, the main reason for the army's failure to uproot the communities in both areas was local Jewish pressure and intercession. That intercession was founded largely on the decades-long history of mutual non-belligerency or outright friendship and collaboration between the neighbouring Jewish and Arab settlements. Without this basic amity or, at least, trust, born in the 1930s and early 1940s, the intercessions of June–July 1948 are inexplicable. (Nor, indeed, is the non-flight of the four villages, earlier, in April–May—as almost all their Arab neighbours decamped into exile—understandable without taking into account that basic foundation of trust in the Jews, in Jewish godwill, and in promised Jewish protection.)

But this local Jewish 'protection' or intercession—by Kiryat Anavim *vis-à-vis* Abu Ghosh and by Zikhron Va'akov and Binyamina *vis-à-vis* Al Fureidis and Khirbet Jisr az Zarka—to be cogent and successful, required national, Cabinet-level endorsement. This was provided, in both cases, by Minority Affairs Minister Bechor Shitrit, whose job was to care for the lot of the country's remaining Arabs, and by the Mapam ministers and officials (Aharon Zisling, Mordechai Bentov, and Moshe Erem). Shitrit, it can be surmised, continuously frustrated, ignored, or contumaciously defied by the army in all that concerned the uprooting of Arab communities during 1948, was probably happy to discover one or two sites still with Arabs and to which he could afford effective protection. The Mapam ministers and officials, likewise, saw in Al Fureidis, Khirbet Jisr az Zarka, and Abu Ghosh (if not Beit Naqquba) good grounds upon which to exercise their moral-ideological fervour. They believed in or, at least, espoused the possibility of, Jewish–Arab coexistence; and here were villages which one could protect at no expense to Mapam or its kibbutzim.

The combination of Shitrit and the Mapam ministers proved effective in frustrating the army's expulsive will; neither the colonels nor Ben-Gurion, even if he was expulsively inclined, could gainsay the past friendliness or collaborationism of the villages in question.

It is worth noting that in both areas the army acted or attempted to act in July in limited defiance of the will of the political leadership. The Cabinet or Cabinet Committee for Abandoned Property may have ruled that Arab villagers were not to be expelled or otherwise moved about without Cabinet-level or the Defence Minister's authorization. But, while there were hostilities, the army continued to go its own way, at least in limited fashion. It could not, perhaps, forcibly expel the inhabitants of Al Fureidis and Khirbet Jisr az Zarka. But it half-heartedly tried to (the tale of the two General Staff / Operations officers) or at least adopted a course that it hoped would have the same evacuative effect. And in the Jerusalem corridor, the army helped nudge, if not actually engineer, the departure of the Abu Ghosh and Beit Naqquba villagers.

Jewish 'protectionism', in both cases, ultimately stymied the army's will. But in the case of Al Fureidis and Khirbet Jisr az Zarka, this 'protectionism' wasn't merely or, indeed, mainly a matter of sentimental goodwill or political morality. Rather, the farmers of Zikhron Ya'akov and Binyamina, to judge by their statements, were primarily driven by economic self-interest. Past Jewish–Arab friendliness or non-belligerency was background; the wine harvest of summer 1948 was pressing foreground. Indeed, at one point the Jewish farmers baldly asked the authorities at least to allow the field-hands to stay for the summer, until the harvest was over, perhaps implying that they would not be troubled overmuch if an expulsion occurred thereafter.

And in the Jerusalem corridor, too, economic interests were at play, at least in all that concerned the relations between Kiryat Anavim and Beit Naqquba. The kibbutz was bent on obtaining its neighbour's removal and lands. Once the villagers decamped, Kiryat Anavim was in the forefront of those opposing Beit Naqquba's return. But the protection afforded Abu Ghosh, both by some elements within the kibbutz (Lisser), and by the politicians, to some extent came to encompass Beit Naqquba's inhabitants or at least those who had ended up in Abu Ghosh in the second half of 1948. (The main village, Abu Ghosh, thus seemingly exercised an indirect protective function over its satellite. The same pattern may also have occurred with Al Fureidis and Khirbet Jisr az Zarka: the Zikhron and Binyamina farmers were mainly exercised over the fate of Al Fureidis; the protection afforded that village was then somehow extended to cover the smaller village, Khirbet Jisr az Zarka.)[58]

[58] It is worth noting that the documents contain more than a trace of a hint that at

In any case, the campaign by Kiryat Anavim and the army against Beit Naqquba was only partially successful: the villagers' return to their former site was blocked, but uprooting those provisionally resettled in Abu Ghosh proved beyond the kibbutz or army's powers. The intervention of the politicians was effective; the cessation of hostilities severely curbed the army's freedom to oust or move about Arab communities. By 1949, military necessity could no longer be compellingly argued.

least some members of Kiryat Anavim through 1948 also desired the permanent removal of Abu Ghosh. Lisser himself apparently in June 1948 sired the idea of transferring its inhabitants to Jaffa. In January 1949, the kibbutz wrote to the authorities in a tone suggesting only reluctant acquiescence in the continued existence of Abu Ghosh as their neighbour. Perhaps here, too, long-term security-mindedness was joined by economic considerations.

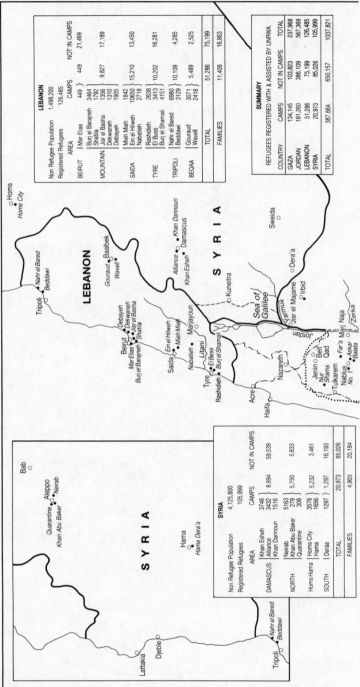

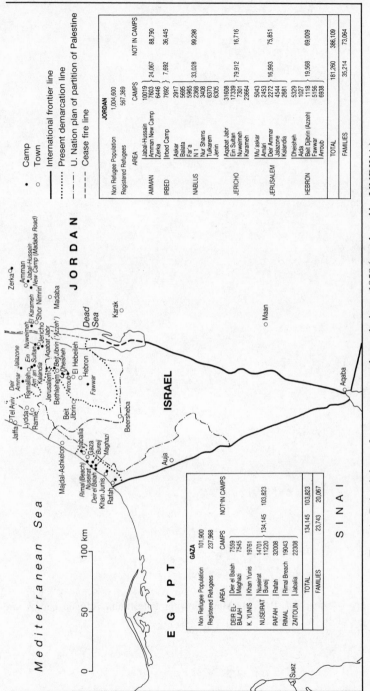

Fig. 4: Palestinian refugee camps and populations, 1958 (produced by UNRWA).

See overleaf for FIG. 4. Map showing Palestinian refugee camps and populations, 1958.

8

The Initial Absorption of the Palestinian Refugees in the Arab Host Countries, 1948–1949

THE FIRST WAVE

The Palestinian Arab exodus began in December 1947, within days of the UN General Assembly passage of the Partition Resolution for Palestine and the start of Arab–Jewish hostilities around the country. The first wave of exiles, who numbered several tens of thousands over December 1947–March 1948, was largely drawn from the urban middle classes—relatively wealthy families from Jaffa, Haifa, and Jerusalem, some with second houses in Lebanon, Nablus, or Amman, who preferred to be out of harm's way for the duration of the conflict. In most cases, they were not compelled to emigrate by Jewish order, pressure, or victory in the field; often, as in Haifa and Jaffa in the first months of the war, the hostilities were relatively minor and these families, if at all, were only remotely threatened. Exile promised enhanced comfort and safety. Few of them thought that exodus would turn into refugeedom, and that their towns, houses, and lands would be conquered and permanently occupied by the Jews; many probably expected the armies of the Arab states to intervene, make short shrift of the Haganah and the emergent Jewish state, and restore them to their towns and property, which, as the months passed, gradually fell under Jewish control.

The arrival in Sidon and Beirut, Ramallah and Nablus, Amman, Cairo, and Alexandria of this first wave of refugees apparently passed without excessive difficulties; the numbers were relatively small, the influx stretched out over months, and the exiles' money oiled the proceedings. There were no major problems of accommodation or sustenance; most had relatives to stay with, second houses, or money to pay for hotel rooms and keep. And the hosts, like the exiles, never conceived of the influx as permanent; within weeks or months, they would be returning to Palestine. It occurred probably to no one on the Arab side at this stage that the war would definitively and radically transform the demography of Palestine to the detriment of the Arab popuiation.

HOSTING THE SECOND WAVE

But things changed dramatically during April and the first half of May, with the Haganah switch to the strategic offensive and the Jewish conquest of most of the Arab-populated areas allotted to the Jewish state in the partition plan. Jewish pressure on the Arab villages of the coastal plain, and the Haganah conquest of parts of Arab Jerusalem and the Jerusalem corridor, Tiberias, Haifa, the Hula Valley and the Galilee panhandle, Jaffa and its environs, Beisan, and Safad, sent some 200,000–300,000 urban and rural Palestinian Arabs fleeing to the safety of the surrounding Arab states (Lebanon, Syria, Egypt, and Transjordan) and the Arab population centres of Gaza, Nablus, Ramallah, and Hebron. These refugees—driven by panic, fear of injury or death in the hostilities, wartime economic privation, and Jewish pressures and expulsion orders—also did not conceive of exile as anything but a temporary expedient or condition. But this second and major wave of refugees, by sheer weight of numbers and chronological concentration, posed a new and radical problem for the Arab host governments, municipalities, and local inhabitants.

Lieutenant-General Gordon MacMillan, the British GOC Palestine, already before the fall of Arab Haifa wrote of the 'suffering' of the Arabs of Samaria (the Jenin–Tulkarm–Nablus–Ramallah area) from the 'enormous influx of refugees from Galilee'. The locals (as well as the refugees) were hit by shortages of food and petrol.[1]

Tens of thousands of the Galilee (and Haifa) Arabs fled to Lebanon, where the influx, in May, was reported by the British Legation to be 'aggravating the shortages existing in certain essential commodities, especially cereals and oil products'. Petrol was in short supply and the Lebanese Government stopped the fuel ration to Palestinian-registered cars; Palestinian car-owners were forced to use the black market. In June, much was made in the Lebanase press of the suffering and death of refugees on the roads to Lebanon

[1] CPV/4/9, 'The General Position in Palestine' by Lt.-General MacMillan, 21 Apr. 1948.

and, according to one French agent (who doubled for Israel), the refugees in Lebanon wanted to end the war at any price 'as they had reached the limit of their endurance'.[2]

The arrival in Egypt during April of what were the first major batches of refugees gave rise to problems, solutions, and attitudes which were soon to characterize most of the host countries. More than 1,100 reached Port Said from Jaffa and Haifa between 25 and 29 April, arriving mostly in 'small steamers, fishing smacks, rowing boats and caiques'. About 270 of these, many of them women and children, arrived in 'open rowing boats, without sails' and without food. The Egyptian authorities placed them in quarantine and Port Said inhabitants took up a collection of funds and food for them.

A 'special quarantine camp' was set up to accommodate some of the refugees at Port Fuad (across the Suez Canal from Port Said), but 'a considerable number' were taken in by local Egyptian families. According to the American Consulate in Port Said, 'many' of the refugees had come with quantities of Palestinian banknotes 'and some of them have been found in possession of narcotic drugs (hashish) which they intended to sell in Egypt'. But there were more serious problems. 'The fact that many of the Arab refugees are of military age and able to bear arms tends to have a dampening effect on the enthusiasm of prospective Egyptian volunteers for military service with the Arab armies. "Why should we go to Palestine to fight while Palestine Arab fighters are deserting the cause by flight to Egypt?" some of them have been heard to remark', reported the consulate. Moreover, 'anti-British feeling' had increased in Port Said as a result of the influx, because the refugees were charging 'that the British are supporting the Jews and neglecting the defence of the Arab population in Palestine'.[3]

The problem of the refugees generating anti-British feelings among Arab host populations was to crop up increasingly through 1948, causing serious concern in Whitehall about Britain's position in the various host countries. Already in February 1948, the High

[2] PRO, CO 537-3986, [Beirut Legation] Summary for the Month of May, 1948; and ISA, FM 2408/16, 'An excerpt from information given by Yosef Sabagh, an agent of the French Consulate in Tiberias and Safad, who had just returned from Lebanon, 24 June 1948.'
[3] NA, Record Group 84, Jerusalem Consulate, Classified Records 1948, 800—Refugees, John P. Robertson, Vice Consul, US Consulate, Port Said, Egypt, to State Department, 29 Apr. 1948.

Commissioner for Palestine, Sir Alan Gordon Cunningham, reported
to London that Transjordan's King Abdullah—with whom he had
lunched that week at Shuneh—had complained about the 'exodus
of Palestine Arabs into Transjordan [saying] . . . they were all arriving
thoroughly anti-British and, hence, might give him trouble'.[4]

Egypt's Moslem religious leaders, meeting as the *Ulema* in Cairo
on 26 April, called on the Arab governments to 'take action' (that
is, to go to war) to liberate Palestine and, meanwhile, 'to give
asylum to Arab refugees from Palestine'. In Egypt, two new camps
were readied for the absorption of refugees, at Kantara and Abbassia,
on the banks of the Suez Canal; most of the 1,600 refugees who had
arrived by sea at Port Said between 2 and 8 May were sent there.
Local Egyptian relief organizations were still supplying food and
clothing, but 'they do not appear to be as enthusiastic now as they
were with the first arrivals'. Quarrels had erupted between the
refugees and the local population. And a rumour had surfaced that
the relief-organization officials intended to visit the camps and
ask the able-bodied among the refugees to join volunteer units to
'fight the Jews. If the men refuse, the Committee [of representatives
of the relief organizations] will request the Government [of Egypt]
to send them back to Palestine.' Some middle-class Egyptians
had reportedly threatened that if the Government did not send
them back, 'they will place them on caiques and send them back
by force . . .'.[5]

Similar sentiments surfaced in Transjordan. 'Some rich [refugees]
have come to Amman and the fact that they are seen frequenting
the banks is causing trouble and ill feeling, the Transjordan folk
seeing no reason why their men should go to Palestine to fight for
the Palestinians, who got their wealth by selling their land to the
Jews', wrote one British observer from Transjordan in mid-June.
She also remarked that money and relief workers were urgently
needed to take care of the mass of poor refugees.[6]

[4] PRO, FO 371–68537 (E3291/4/31), Sir A. Cunningham to J. M. Martin,
Colonial Office, 2 Feb. 1948.

[5] PRO, FO 371–68371 (E5528/11/65), Sir Ronald Campbell, Cairo (British
Middle East Office), to Foreign Office, 30 Apr. 1948; and NA, Records Group 84,
Jerusalem Consulate General, Classified Records 1948, 800—Refugees, Philip
Ernst, US Consul, Port Said, to US Embassy, Cairo, 13 May 1948.

[6] Jerusalem and East Mission Papers (JE&EM), Box LXXIII/1, Winifred
A. Coates, El Husn, Transjordan, to 'Mabel' (Jerusalem), 15 June 1948.

Palestinian leaders tried to combat this ill-feeling towards the refugees. Suleiman Tukan, a pro-Hashemite dignitary from Nablus, on 11 June on Amman Radio comprehensively defended the refugees by charging that the Arab states were 'responsible' for the plight of the Palestinians as they had not heeded the Palestinians' cries for help. The essence of Tukan's defence was that the Arabs had not fled the country of their own free will. Be that as it may, Tukan appealed to the refugees of the Samaria and Hebron districts to return to their homes; the appeal may have been the price Tukan had had to pay the Transjordanian authorities for allowing him to broadcast his 'defence' of the refugees.

The start of the First Truce between Israel and the Arab States on 11 June also apparently spurred the Syrians to try to 'push' their refugees back across the border, back to the abandoned Arab villages in the (Galilee Panhandle) Hula Valley. Israeli intelligence assessed that Damascus had adopted this policy 'to remove from itself the economic and organisational burden [of their upkeep]' and 'to introduce a Fifth Column [back] into the Jewish areas cleared [of Arabs]'. Meanwhile, the Syrians were providing at least some of the refugees with food supplies.[7]

The IDF, of course, had received orders—stemming from both tactical military and strategic-political considerations—to halt with fire the return of refugees back into Israeli-held territory.[8] Local IDF intelligence officers in various sectors of the country believed that Arab refugees were infiltrating back into Israeli-held territory mainly to harvest the summer crops rather than to permanently resettle in their former villages. The refugees were driven by 'real hunger'. The situation of the refugees along the Israeli–Syrian border was so desperate that they braved IDF bullets to foray into the fields in the Galilee panhandle during the nights. Some refugee farmers even returned at night to water their fields 'in the hope that they will manage to plant vegetables on [for?] their return to their village with the [start of the] truce'. IDF Intelligence warned that the process, if allowed to go on, would eventually result in the

---

[7] KMA-PA 100/MemVavDalet/1—9, *Batziburiut Ha'Aravit* (in the Arab public), Foreign Ministry Middle East Affairs Department, 11 June 1948. Copy also in ISA, FM 2570/6; and ISA, FM 2408/16, 'An Excerpt from Information Given by Yosef Sabagh . . .'.

[8] See Essay 6 in the present volume.

refugees indeed permanently resettling in Israel.[9] The IDF duly barred the refugees from returning to Israeli-held territory.

The majority of the 300,000 refugees who had left Israeli-held territory by the 15 May invasion of Palestine by the Arab armies fell under the control, and became the immediate responsibility, of King Abdullah and his government; his army, the Arab Legion, had occupied most of the Nablus–Ramallah–Jericho–East Jerusalem– Hebron area. Sir Alec Seath Kirkbride, the British Minister to Amman, grasped the significance of the events and sensed what was to come in terms of the refugees. 'It is not even possible to foresee how the Transjordan authorities are to provide food and shelter during the coming winter' for the thousands 'now living in destitution,' he wrote to Foreign Secretary Ernest Bevin. The refugees themselves, he added, were 'becoming increasingly vocal' in demanding an end to the war 'at all costs'. Their demand, he thought, was 'based on the assumption that if hostilities cease, they will be able to return to their homes in the Jewish areas. [But] why the Jews should permit them to do so, now that events have solved the most difficult problem which originally faced the Jewish state, that of a large Arab minority, it is difficult to see. In fact, the Arab world is likely to be faced with the resettlement of these refugees and not with their repatriation . . .' The Arab leaders had never 'anticipated' the mass flight of the Arab population, 'often without there being any imminent threat', from their homes in the areas allocated to the Jews under the partition plan, he added as an afterthought.

Kirkbride's letter to Bevin sparked the first major internal Foreign Office debate about what could be done for or about the Palestinian refugees. Lance Thirkell thought that the International Refugee Organization, which, he said, had 'done so much to help Jews to get into Palestine', could help the 'unfortunate Arabs who have been

---

[9] KMA-PA 100/MemVavDalet/1—9, *Batziburiut Ha'Aravit*, Foreign Ministry Middle East Affairs Department, 11 June 1948; and ISA, FM 2570/6, 'Tsur' (the code name of a local intelligence officer) to Haganah Intelligence Service, 7 June 1948. The IDF duly barred the refugees from returning to Israeli-held territory.

pushed out incidentally. Singularly little has been heard of the fact that the Jewish incursion into Palestine has produced an exodus of 300,000 refugees. I should have thought that this would have had some [anti-Israeli] propaganda value.' David Balfour also thought the IRO could help. But A. W. Wilkinson minuted: 'The IRO has insufficient funds to take on any more refugees and doesn't deal with "internally displaced" people.' She rejected Thirkell's and Balfour's comments that the IRO had helped Jews get to Palestine and said that, in fact, the organization had assisted in the resettlement of European Moslem refugees. H. Beeley commented that in Palestine the issue was not really one of 'internal displacement'.[10]

The '10-days fighting' (between the end of the First Truce and the beginning of the Second Truce) in mid-July both highlighted and exacerbated the plight of the refugees. A further 100,000 Palestinians joined the ranks of the exiles, the bulk of them expelled by the IDF from Lydda and Ramle to the Transjordanian-held part of Palestine (Latrun–Ramallah–Qalqilya–Jericho). The massive, concentrated influx (on top of the defeat administered to the Arab Legion) shook Abdullah's kingdom. The British colony in Amman considered evacuation, but rejected the idea after Legion commander General John Glubb said he was 'not pessimistic' about keeping order in the city: 'it may be possible to keep Amman Suk [market-place] and Palestinian refugee elements under control'. Abdullah and Glubb wanted to 'end hostilities' but were 'frightened' of the reaction of the other Arab states and of 'public feeling'. There had been several anti-British and, implicity, anti-Abdullah demonstrations in Amman and elsewhere in the country.[11]

Transjordan's refugee problem had 'become acute' and the government was unable to cope. A rough idea of the situation in July in Ramallah, a town of about 10,000 and the first major way-station of the Lydda–Ramle refugees, is provided in a message from the municipality to King Abdullah, transmitted via the Arab Legion's radio communications system and monitored by Israeli intelligence: 'Seventy thousand people are scattered in the streets, the great

[10] PRO, FO 371–68570 (E9239/4/31), Sir A. Kirkbride, Amman, to E. Bevin, Foreign Office, 2 July 1948 and attached minutes by FO officials.
[11] PRO, FO 371–68575 (E9723/4/31), C. M. Pirie Gordon, Amman, to Foreign Office, 18 July 1948; and NA 501 BB. Palestine/7–1548, U.S. Legation, Amman to Secretary of State, 15 July 1948.

majority of them impoverished [and] suffering from a major lack of
basic goods and water, constituting a serious health hazard', reported
acting mayor Hanna Khalaf. 'The municipal council beseeches Your
Highness to issue [an order] to the city of Ramallah to evict [the
refugees] as the city cannot cope.' Abdullah answered unsympathe-
tically that the city must display 'patience towards your brothers',
and that relief would soon be on its way. A graphic description of
the situation in the town was provided by an English-language
Ramallah radio broadcast:

The smell is beginning to be bad, in so many places [said the reporter-
observer], and what can the girls do with small brooms of twigs, in
classrooms now holding three hundred people? Look at the dignified
families in the [olive] groves and rooms, attempting to make privacy for
themselves with stretched out blankets and dresses . . . Look at the women
washing clothes and baking some precopis flour for their hungry families
. . . There won't be a drop of flour left in Ramallah in three days . . . There
are seventy thousand people to feed . . . Transjordan rushed over dates,
flour and bread, but it is a drop in the ocean . . .

Seen from Tel Aviv, the refugees constituted a major burden for
the Transjordanian authorities, particularly the military, clogging
roads, requiring food and shelter, diverting Arab Legion resources
and energies, as well as undermining the morale of the local
population in the towns north and west of Jerusalem.[12]

Towards the end of July thousands of Lydda–Ramle refugees
moved from Ramallah, eastwards, into Transjordan, to Amman.
There, too, most were housed in school buildings (the schools were
recessed for the summer vacation) or camped out in olive groves
and private gardens. One British observer, Winifred Coate, a
school headmistress, found them 'lying on old sacks and rags',
destitute. They lacked bedding and cooking utensils. 'Life out of
doors like this does not seem unnatural to many of them who were
accustomed to go out and camp in their vineyards at this time of

[12] NA 501 BB. Palestine/7–1548, Amman to Secretary of State, 15 July 1948;
ISA, FM 2569/13, 'From Monitoring the Legion Wavelength', 21 July 1948; PRO,
FO 371–68578 (E10440/4/31), Sir H. Dow, British Consul-General, Jerusalem, to
Foreign Office, 19 July 1948, enclosing text of Ramallah Radio broadcast, undated;
and ISA, FM 2569/13, the Research Department (the Foreign Ministry's intelligence
department), to Y. Shimoni, acting director of the Foreign Ministry Middle East
Affairs Department, 19 July 1948.

year, but this is in the middle of Amman and is most unsuitable in a town', she commented.[13]

The US vice-consul in Amman, Wells Stabler, towards the end of July reported that the Transjordanian estimate of Palestinian refugees in Transjordan (east of the Jordan River—in Amman, As-Salt, Zarka, and Irbid) stood at '80,000–100,000', and Abdullah had felt obliged to close the kingdom's frontiers to additional refugees. The British estimate at this time, based on figures provided by Glubb, was about 56,000 refugees in Transjordan proper and some 70,000–100,000 in the areas west of the Jordan River occupied by the Arab Legion.[14]

How were the Transjordanian authorities coping with the refugees? Coate thought that 'the people of T.J.[Transjordan] had risen to the occasion rather better than might have been expected, as they have no experience of the kind of organisation needed'. In Amman, two ladies' welfare societies had raised funds. In Zarqa, a 'Ladies Committee' had doled out to each family 'a small packet of rice and sugar and tea and a piece of soap; also the most destitute had been given one blanket per family'. But this order of voluntary assistance, as Coate realized, was 'only a drop in the bucket of the great need'.

Coate, friendly with the Transjordanian finance minister's wife, appears to have been well-informed, at least as regards the refugees east of the river. They were getting, at the end of July, two small loaves of bread per head per day from the government. In addition, pregnant women and some children were getting milk. The government planned, she said, to hand out 300 mils (about six shillings) per month per head to cover other necessities; but she probably treated this information sceptically in light of Transjordan's poverty.[15]

The scepticism of the British consul in Amman, Pirie Gordon, was more comprehensive and far-seeing. 'I doubt whether Transjordan Government is capable of preparing any plans for organizing the

[13] J&EM LXXIII/1, Winifred Coate, principal, CMS Girls' School, Amman, to 'Mabel' Jerusalem, 30 July 1948; and ISA, FM 2569/13, 'Hiram' to Foreign Ministry Research Department, 19 July 1948.

[14] NA, Record Group 84, Jerusalem Consulate General, Classified Records 1948, 800—Refugees, Amman (Stabler) to State Department, 26 July 1948; and PRO, FO 371–68576 (E10219/10006/4), Amman (C. M. Pirie Gordon) to Foreign Office, 29 July 1948.

[15] J&EM LXXIII/1, Winifred Coate, Amman, to 'Mabel' Jerusalem, 30 July 1948.

future relief, and in any case the main difficulty lies in the burden imposed on the country's minute financial resources', he wrote. The local authorities foresaw starvation and the outbreak of epidemics if nothing substantial was done, quickly. And he was not particularly impressed by the level of public generosity displayed. In one Trans-jordan district, with 16,000 refugees, he wrote, the local authorities had managed to raise by public subscription only P£1,800. 'In Amman suggestion that merchants should contribute has been flatly rejected and an offer of funds of the Supreme Moslem Council and of the Ramle municipality approximately a total of P£90,000 has so far come to nothing because the Arab banks who hold these accounts prefers [*sic*] to have no funds available.' Meanwhile, prices had shot up, and the refugees themselves had quickly run out of savings and were begging or hungry; 'there has been a marked increase in crimes against property', Pirie Gordon added.

In short,

being idle, homeless, bereft of all their possessions and discontented with the relief efforts made on their behalf[,] the refugee communities in Amman and Salt at least, are fertile breeding grounds for embittered and opposed sentiments . . . Britain is accused as usual of being mainly responsible for their plight. The Arab rulers are held guilty of misleading them and involving them in the loss of their farms and settled lives. Local authorities regard them as a menace to the peace and security of the country and believe that communists are at work amongst them agitating against established order here,

wrote Pirie Gordon.[16]

In the Transjordanian-occupied part of Palestine, the situation was no better. Sir Hugh Dow, the British consul-general in Jerusalem, reported that 'local efforts to deal with the problem are largely uncoordinated and limited to bare emergency measures by local authorities and social committees . . . The main requirements of Arab refugees still inside Palestine are food, water, blankets and clothing. As the autumn approaches, the need for the two last named will increase and also for temporary or permanent housing . . .' Dow was thoroughly disgusted with the Arab governments' attitude to the refugees: they appeared less concerned with providing relief than getting rid of the 'refugee incubus'.

[16] PRO, FO 371–68576 (E10219/10006/4), Amman (Pirie Gordon) to Foreign Office, 29 July 1948.

Dow suggested that the 'best' solution for the problem of the refugees from the Jewish areas would be their resettlement in Arab countries. Israel would pay for the abandoned Arab lands and other property in Palestine, and the money would go towards compensating individual refugees and towards 'the resettlement of the poorer refugees'. At the same time, fund-raising in Britain ('a Lord Mayor's Fund') could cover the cost of repatriating Palestinians to Arab or Jewish-held areas of Palestine ('though they should not be encouraged to do so', commented Dow). Dow said that the Lord Mayor's Fund should be administered by Britain rather than the Arabs (a suggestion with which Pirie Gordon concurred). He added, however, that one should expect Arab League opposition to the scheme 'since it will have a purely humanitarian and practical object and will not fit in with their political desire to rid themselves of refugees in their territories in the first place . . . and to send large Arab populations back into Jewish areas as a nucleus for future trouble'.[17]

Dow was reacting to a cable from W. E. Houstoun-Boswall, the British Minister in Beirut, who on 21 July had proposed the establishment of a Lord Mayor's fund as 'a spontaneous act on the part of the British public to show their sympathy and friendly feelings for the Arabs' and to counter the trend in the Arab countries of vilification of Britain as a betrayer of the Arab cause. Houstoun-Boswall, like many other British diplomats in the Middle East at the time, was at best dismissive of the Palestinians, who had chosen 'to run away rather than fight'; by their 'general exodus', he had written, the Palestinians had 'shown themselves to be rather contemptible and perhaps not so deserving of sympathy . . .'. But Britain must act charitably, to show that the British remain the Arabs' 'best friends'.[18]

The American assessment of the state of the refugees west of the Jordan River was at first somewhat more sanguine—'Condition . . . not yet desperate'; it forecast complete destitution 'shortly', followed by 'hardship and danger to health' with the onset of winter. But an on-the-spot inspection by consular officials a fortnight later, in mid-August, produced a far darker picture. The Ramallah area

[17] PRO, FO 371–68576 (E10235/4/31), H. Dow, Jerusalem, to Foreign Office, 29 July 1948 and (E10219/10006/4), Amman (Pirie Gordon) to Foreign Office, 29 July 1948.
[18] PRO, FO 371–68575 (E9992/4/31), Houstoun-Boswall, Beirut, to Foreign Office, 21 July 1948.

was flooded with some 100,000 refugees, or almost twice the number of the normal local population. 'Condition refugees appalling . . . Live along sides road under trees . . . Housing facilities or even tents not available . . . Nights become very cold mid-September and rains start soon after . . . Water supply . . . precarious . . . water probably contaminated . . . Definitely possible that water supply may give out completely before end August.' The Transjordanian Government was daily distributing 250 grams of bread per day per refugee (half the amount east of the Jordan). The doctor accompanying the consular officials ruled that this represented about 600 calories per day and was 'insufficient [to] sustain life for long'. Moreover, 'sanitation practically nonexistent . . . no hospital beds available', little vaccine left for immunization against typhoid, and none at all against typhus, smallpox, diphtheria, and cholera. While 'good weather and good luck', to date, had left the adult death-rate normal, and only twenty suspected cases of typhoid, one of diphtheria, and one of meningitis had been reported, the worst was to be expected with respect to health during the following weeks. The consul general summed up the situation thus: 'Complete lack of organization apparent. Local authorities overwhelmed by magnitude problem and admit own inability cope with situation. No funds or qualified personnel available for organizing camps, food distribution, sanitation and immunization programs etc.'[19]

The situation was no better north of Ramallah. Major Hackett-Paine, a former British liaison officer with the Transjordanians, in mid-August toured the Iraqi-controlled Samaria District (the Nablus–Tulkarm–Jenin 'triangle') and found the state of the refugees—who he was told numbered in the District between 100,000 and 120,000—'very bad' and their morale 'very low'. Most lived in the open, under trees, in caves, and in unfinished houses; few had blankets and there was insufficient food. A major health problem was expected when the rain came. Most blamed the British for their plight 'and their hatred for us plus America and the Arab League is unbelievable'. As well, since the fall of Lydda and Ramle, 'Abdulla's name is just plain dirt'. Hackett-Paine, too, pleaded for British relief aid (to be channelled not through the Arab League).

[19] NA Record Group 84, Jerusalem Consulate General, Classified Records, 1948, 800—Refugees, John MacDonald, US Consul-General, Jerusalem, to Secretary of State, 27 July 1948 and John MacDonald to Secretary of State, 12 Aug. 1948.

The Foreign Office's Lance Thirkell thought the major's report 'ungrammatical and over-emotional' but that it conveyed well the 'acute misery' of the refugees in Samaria. Another Foreign Office official minuted: 'We may be hated now but that is nothing like the hatred which the Jews are laying up for themselves in the future if they don't allow these people back.'[20]

The great majority of the Palestinian refugees at this time—as in future decades—were concentrated in Transjordan and in the Transjordanian-held eastern areas of Palestine; smaller groups had moved to Egypt, Lebanon, Syria, and Iraq.

By August, according to Jefferson Patterson, the US chargé d'affaires, there were some 14,000 refugees in Egypt, dispersed in private homes and in the makeshift camps along the Suez Canal and in Cairo. 'All are reported to be in great need', he reported. He had been informed that Egypt was considering expelling its Jewish community and confiscating their property 'in order to provide aid and houses for the Palestine Arab refugees'; Egyptian officials denied this. The Egyptian press, he reported, was 'indignant' at the world's silence about the plight of the refugees. There were another 50,000 refugees in Egyptian-occupied Palestine, mainly concentrated in the Gaza area.[21]

In Lebanon, mainly concentrated in camps outside Sidon, near Lake Kar'oun in the eastern Bek'a Valley and at Bint Jbail, just north of the Israeli border, were some 50,000 refugees. The government was providing each with 10 kilograms of flour and three Lebanese pounds per person. Part of this was covered by a grant to the Lebanese Government of E£100,000 from the Arab League. The Lebanese were pressing UN Mediator for Palestine Count Folke Bernadotte to obtain Israeli permission for their repatriation. The Lebanese Christians viewed this massive influx of Moslems as a serious threat and 'would resist any attempt to [permanently] resettle refugees in Lebanon,' reported the British Minister, Houstoun-Boswall. He agreed with Dow's assessment that the solution to the refugee problem would ultimately have to be resettlement in the

---

[20] PRO, FO 371–68677 (E11504/10748/31), copy of Hackett-Paine's report covering note from HM Consulate General, Jerusalem, to Eastern Department, Foreign Office, 25 Aug. 1948, and minutes by several FO officials.

[21] NA, Record Group 34, Jerusalem Consulate General, Classified Records 1948, 800—Refugees, Patterson (Cairo) to Secretary of State, 7 Aug. 1948.

Arab countries, 'but Lebanon for obvious reasons would have to be excluded from any such scheme', he thought.[22]

By the end of July, only some 200 Palestinian refugees had reached Iraq. But Iraq, according to its Prime Minister, was about to receive another 5,000 (mainly from the 'little triangle' of Jaba, 'Ein Ghazal, and Ijzim, an originally Iraqi-controlled Arab enclave in Palestine's coastal plain conquered by the Israel Defence Forces during the Second Truce). The Iraqis, according to the British Legation in Baghdad, had made no provision for the refugees and had raised no public funds on their behalf.

Another 70,000 refugees were camped out, in poor condition, in Syria, mostly along the border with Israel. Both Iraq and Syria were already being targeted by senior British diplomats in the Middle East (as by Israeli officials) as the most appropriate sites for the permanent resettlement of most of the refugees. As a senior official in the British Middle East Office in Cairo put it, 'in Iraq and Syria . . . there are sufficient latent resources . . . to support a population several times the present numbers. In fact, in Iraq rapid progress in development is to a considerable extent dependent on an increase in population.' Of course, such argumentation was presented most forthrightly by British representatives in precisely those countries —such as Egypt and Lebanon—on which the burden of the refugee influx, be it for political, economic, or demographic reasons, weighed most heavily.[23]

Two major factors emerged from the cursory British and American diplomatic investigations in late July and early August 1948 of the Palestinian refugee situation. The conditions of their existence were, by and large, 'appalling'—'worse than anything he had ever seen', according to Bernadotte—and most of the Arab states were doing little, if anything, for their relief. As the Transjordanian Prime Minister told Kirkbride, Abdullah's appeal to his fellow Arab leaders for aid had secured only 'unfulfilled promises of money from Saudi Arabia and the Yemen'. Abdullah and his Prime Minister agreed, according to Kirkbride, that 'some of the Arab

[22] PRO, FO 371–68576 (E10232/4/31), Houstoun-Boswall (Beirut) to Foreign Office, 28 July 1948.
[23] PRO, FO 371–68576 (E10234/4/31), Richmond (Baghdad) to Foreign Office, 31 July 1948; (E10235/4/31), minute by Lance Thirkell, 4 Aug. 1948; and FO 371–68578 (E10456/4/31), BMEO (Cairo) to Foreign Office, 3 Aug. 1948.

states best able to bear a heavy burden . . . are doing the least of all . . .'.[24]

THE START OF INTERNATIONAL RELIEF EFFORTS

Something obviously had to be done. British Foreign Secretary Bevin felt that the key lay in the hands of the Arab states themselves. On 13 August he sent a forceful cable to Britain's posts in the Middle East stating: 'I am by no means convinced that the Arab States, and particularly those who can best afford it, are using their resources to the best advantage for the relief of Arab refugees.' He instructed the British diplomats to inform the Arab governments that 'more could and should be done by the Arab States themselves. His Majesty's Government realise that the main concern of the Arab States is that the refugees should be allowed to return to their homes and that they wish to do nothing to prejudice this possibility.' But even if this became possible, much time would elapse before implementation could conceivably take place; meanwhile the refugees were living in misery. Bevin particularly feared the outbreak of major epidemics. He called on the Arab states to supply funds and equipment for the relief of the refugees, a course which might persuade the powers outside the Middle East also to come forward with assistance; and he proposed that a local, Middle Eastern refugee relief organization be set up, to co-ordinate the efforts on behalf of the refugees.[25]

A similar feeling surfaced during July, August, and early September among Bernadotte and his staff, as it gradually became clear to them that Israel had no intention of allowing a return of the refugees to their homes. Bernadotte had the UN secretary general send Sir Raphael Cilento, the Australian director of the Division of Social Activities in the UN's Department of Social Affairs, to investigate the situation of the refugees. Cilento toured Transjordan and in early August reported on their condition. Bernadotte subsequently formulated a three-stage plan for refugee relief: (1) Immediate

[24] PRO, FO 371–68677 (E1088/4/31), Chapman Andrews (BMEO, Cairo) to Foreign Office, 16 Aug. 1948; and FO 371–68677 (E11025/10748/31), A. Kirkbride (Amman) to Foreign Office, 18 Aug. 1948.
[25] PRO, FO 371–68578, Foreign Office to Cairo, Baghdad, Beirut, Jidda, Damascus, Amman, BMEO, etc., 13 Aug. 1948.

relief of absolute basic needs, (2) relief from September to December, in conjunction with a thorough study of the problem, and (3) a long-range programme, covering January–September 1949. At the same time, he began personally to solicit assistance for the immediate relief of the refugees from dozens of governments and international bodies, asking for funds from the wealthier states, cereals from cereal exporters, and so on. Within his administration, a Disaster Relief Project (later called Refugee Relief Project) was set up, headed by Cilento, to co-ordinate these efforts and to oversee the distribution of the contributions. But Bernadotte was assassinated in Jerusalem by Jewish terrorists on 17 September, and Cilento's organization proved substantially ineffective, given the lack of any major contributions from the wealthier Western states and the lack of effective organization and generosity by most of the Arab states. Britain, for example, had supplied thousands of tents. For weeks these remained unopened and unused in a warehouse in Beirut, where the major relief-supplies bottleneck existed, due to a mixture of corruption and inefficiency. Money was not forthcoming and basic foods, water, medicines, doctors and medical facilities, tentage, clothing, and bedding all remained in very short supply or nonexistent through August and September in most of the refugee communities.

In late September an official of the International Committee of the Red Cross toured the refugee centres in Lebanon, Syria, Transjordan, and Palestine and reported that, despite the 'hullabaloo' surrounding international contributions, the 'tragic fact is that substantially nothing in food or goods have reached refugees up to October 1' apart from the direct aid proffered by the Arab governments and local communities. Stanton Griffis, the US Ambassador in Cairo, found the situation 'extremely delicate' in view of the almost complete absence of any US government or private contributions. Such aid was 'vital', said the ambassador, as the refugee issue had moved to the top of the Arab agenda in negotiations about a possible solution to the Middle East conflict.[26] Griffis felt very strongly about the refugees, not only within the context of American vulnerability and US interests in the Middle East. In December 1948, he was appointed by the UN secretary general the first

[26] NA 501. BB Palestine/10–648, Cairo (Griffis) to Secretary of State, 6 Oct. 1948.

director of UN Relief for Palestine Refugees in the Near East, effectively replacing Cilento. As his adviser, Griffis appointed Dr Bayard Dodge (former president of the American University of Beirut).

A similar message reached the State Department from Amman, where Stabler reported that, with the onset of rain, the condition of the some '200,000' refugees in Transjordan proper and in Transjordanian-occupied Palestine was 'severely deteriorating'. Transjordanian resources were too meagre to cope and, thus far, the only relief supplies to have arrived through the UN was a 'small amount [of] powdered milk for children and pregnant women . . . Immediate relief is not only of importance from humanitarian standpoint, but also as essential factor in arriving at satisfactory solution Palestine problem based on Bernadotte's conclusions [in his Interim Report and 'plan' of mid-September]. This running sore of refugees will make Arab acquiescence therein [that is, in the plan] more difficult and odious as it will remain evidence of UN inability cope with complicated yet urgent problem.' Stabler warned that further delay in the provision of relief 'can only result in appalling number of deaths'. He strongly urged the State Department to pressure the UN to begin providing the necessary supplies.[27]

The sense of urgency and pressure from the field culminated, for the US Government, with the cable of 17 October to the President and the Secretary of State from the new US Special Representative to Israel, James McDonald. Of the 'approximately 400,000 refugees', he wrote, an estimated '100,000 old men, women and children who are shelterless and have little or no food' would die when the rains came. The refugee problem was reaching 'catastrophic proportions and should be treated as a disaster'. The UN relief machinery was 'both inappropriate and inadequate' and had led to 'gross inefficiency and wastefulness'. Immediate action, guided by a comprehensive programme of relief, was needed to 'avert horrifying losses'. The present system of getting the relief from the contributor through the UN machinery to the countries and then the individual refugees was a failure, with wastage *en route* estimated at as high as '90 per cent'. McDonald recommended the immediate transfer of the whole relief operation to the hands of the Red Cross. McDonald believed the

[27] NA 501. BB Palestine/10–148, Amman (Stabler) to Secretary of State, 10 Oct. 1948.

UN administration, Israel, and the Arab states could be persuaded that such a change would be in everybody's interest.[28]

The obvious anarchy in the relief efforts and the growing suffering of the refugees, issuing in increased international (mainly American) pressures, prompted the UN General Assembly resolution of December 1948 setting up the United Nations Relief for Palestine Refugees in the Near East (UNRPR). The UNRPR co-ordinated with and acted through the International Committee of the Red Cross, the League of Red Cross Societies, and the American Friends Service Committee in the distribution of the relief supplies. The Red Cross was given 'jurisdiction' over the refugees in Israel and the Transjordanian-occupied parts of Palestine; the League of Red Cross Societies, over Transjordan, Syria, and Lebanon; and the American Friends Service Committee (Quakers), over the Egyptian-occupied Gaza District. The UNRPR effectively replaced the Disaster Relief Project. In turn, in December 1949 the UNRPR was succeeded by the United Nations Relief and Works Agency for Palestine Refugees in the Near East (UNRWA), which took over and ran the refugee camps during the following decades.[29]

HOSTING THE FOURTH WAVE, THE WINTER OF 1948–1949 AND ORGANIZAING THE FIRST CAMPS

The refugee situation took a major turn for the worse towards the end of 1948, with the successful Israeli offensives of October and December 1948–January 1949, resulting in the flight of a further 150,000–200,000 Palestinians, most of them to the Gaza area.

In early November 1948, the Gaza Strip turned almost overnight into the worst hit district of Palestine. Some 210,000–230,000 refugees from the southern coastal plain and the northern Negev now packed an area (roughly 25 miles long by 5 miles wide) previously inhabited only by 60,000 local inhabitants. The first weeks were chaotic. The bulk of the refugees, after a lengthy trek from the north or east, came to rest 'under trees or just along the road . . .', reported UNICEF official Dr P. Descoeudres, who

---

[28] NA 501. BB Palestine/10–1748, Tel Aviv (McDonald) to Secretary of State (and President), 17 Oct. 1948.

[29] See Don Peretz, *Israel and the Palestine Arabs* (The Middle East Institute, Washington, DC 1958), 8–13.

visited the Strip on 11–12 November. One observer, who visited the Strip between 9 and 11 December, reported that in and around Gaza town now lived 60,000 refugees in addition to the original population of 25,000. 'They pack sidewalks, take up the vacant lots and the public market, occupy barnyards, and generally seem to fill in every empty space . . . They live in churches, mosques, schools . . .'.[30]

But already, a small proportion of the refugees—probably directed by Egyptian officers—had taken up residence in camps. By mid-November some 20,000 refugees had moved into the former British Army camps at Bureij and Nuseirat, south of Gaza, filling the ramshackle buildings and tents set up alongside the buildings in the camp compound. During a visit to Bureij, one observer described the refugees ensconced in the camp's former mess hall: 'It was about 50 feet wide and 120 feet long. There must have been 800 people in the place. They had staked out little cubicles for themselves using rags or flattened gasoline tins to make the borders between the family groups. Everyone was very dirty and cold. In one cubicle we saw a group of ten people . . . looking at an old woman on the floor who had just died.'[31] At Nuseirat camp, part of which had served in Mandate days as a military prison, families of seven or nine persons lived in each 'one-prisoner cell', six by eight feet in size. The original roofs had vanished, but the Egyptian Army had installed makeshift canvas and tin roofing 'which does not keep out the cold but does help lessen the amount of rain entering each room'. The wooden covers installed in the central latrines were immediately removed by the refugees and used for firewood.

During the following weeks the refugees, initially diffused, clustered together in new concentrations, which gradually became, under Egyptian Army and UNRPR management, organized camps. Altogether, eight large organized camps emerged in the Strip (Jabalia, Beach, Nuseirat, Bureij, Maghazi, Deir al Balah, Khan Yunis, and Rafah). In the new refugee camp, outside Khan Yunis, south of Bureij, with a 'supposed 40,000' population, the refugees lived

[30] American Friends Service Committee Archive (Philadelphia) (henceforward AFSC), Foreign Service 1948, Palestine, 'Report of Visit by Dr Descoeudres to Southern Palestine, Nov. 11–12 1948', 13 Nov. 1948, Beirut, and 'Confidential Memorandum', John Devine to Ambassador Stanton Griffis (Cairo), 13 Dec. 1948.
[31] Ibid.; ibid.

initially fifteen to twenty per tent in tents originally intended for seven soldiers. There was a major shortage of tents, which was to plague the Gaza refugees at least until February 1949.[32] In Rafah, with some 30,000 refugees in mid-December, only 4,000 lived in a tent-camp, with 'the rest huddled wherever they can find any kind of shelter'.[33] The problem of shelter sorted itself out over December 1948–January 1949, as the bottlenecks in Beirut were unclogged and tentage began to flow to the Strip. By the start of March 1949, some two months after the Quakers took over the refugee relief operations in the Gaza Strip, a senior American Friends Service Committee (AFSC) figure was able to report that 'all refugees [in the Strip] are now under some kind of shelter, mostly tents'. But still more tents were needed, to reduce the overcrowding.[34]

In the early November 1948 days in the Gaza Strip, the refugees endured appalling conditions. Apart from shelter, the supply of food, water, medical services, and clothing were the primary problems. One American Red Cross official, R. T. Schaeffer, reported that 'thousands of refugees have fallen by the roadside from starvation' and there was widespread dysentery. The Egyptian Army, recovering from the series of defeats and routs that ended on 9 November, barely had time or resources for refugee relief. A UN official, F. G. Beard, reported that the Egyptian Army and the Cairo-based Arab Higher Refugee Council had been 'grossly negligent'.[35] The Egyptians initially, in early November, distributed some flour and rice to the refugees. The Strip's military governor said he had received E£12,000 which he had used for this purpose. But by mid-November the supplies had apparently run out and the refugees were receiving no food rations from the Egyptian authorities, according to foreign observers on the scene. Indeed, all the observers reported that, the moment they reached a refugee encampment or

---

[32] AFSC, Foreign Service 1948, Palestine, 'Memorandum', Gillespie Evans to Ambassador Stanton Griffis (Cairo), 26 Dec. 1948; 'Confidential Memorandum', John Devine to Ambassador Stanton Griffis (Cairo), 13 Dec. 1948; and 'Memorandum'. Gillespie Evans to Ambassador Stanton Griffis (Cairo), 26 Dec. 1948.

[33] AFSC, Foreign Service 1948, Palestine, 'Confidential Memorandum', John Devine to Ambassador Stanton Griffis (Cairo), 13 Dec. 1948.

[34] AFSC, Foreign Service 1949, Palestine, 'Report of Clarence E. and Lilly Picket on their Visit to the Middle East', 3 Mar. 1949.

[35] NA 501. BB Palestine/11–1648, Cairo (Patterson) to Secretary of State, 16 Nov. 1948; and 501. BB Palestine/12–748, Cairo (Patterson) to Secretary of State, 7 Dec. 1948.

concentration, 'we were surrounded by crowds of mothers bringing their children and begging for food, and adolescents and men, threatening with doubled fists . . . stating they were starving'; the observers were 'smothered . . . with entreaties for food'. According to the International Red Cross doctor who treated the Gaza refugee population, 'at least ten children died from starvation each day in the camps' in mid-November. There was no milk ration for children.[36]

A month later, in mid-December, the situation had somewhat improved. In Bureij camp, with some 13,500 refugees, each person was receiving two and a half kilograms of flour per ten days (the Arab Higher Refugee Council having sent 1,000 tons and Australia, via th UN, another 500 tons). 'There are no prostrate people visible such as one read about in the famines in India and China,' reported one observer. But of the 10,000-odd refugees clustered in the Khan Yunis area, 'no more than half . . . [were] receiving the flour ration. The rest [were] begging for Egyptian Army leftovers, stealing oranges, or spending whatever money they have brought with them.' The Egyptian military governor in the Khan Yunis area 'seems to have very little interest in the problem', reported one western observer.[37]

But by the end of the month, the food situation seems to have stabilized, with most or nearly all the refugees receiving regular food rations. Most of the refugees had been issued with Egyptian identity cards and each card-holder received two and a half kilograms of flour per ten days (roughly half a pound of bread per person per day). Bakeries had been set up, with the refugees bringing in unbaked loaves and paying the baker one piastre (1,000 piastres to an E£) per twenty loaves. In the Gaza area, Egyptian soldiers apparently routinely gave away to the refugees one loaf of bread of the four they drew as their daily food allowance. The Egyptians, funded by various UN agencies and private contributions, maintained this level of nutrition of the refugees (about 600 calories per day) until January–February 1949 when, under the

---

[36] AFSC, Foreign Service 1948, Palestine, 'Report of a Visit by Dr Descoeudres to Southern Palestine, Nov. 11–12, 1948', 13 Nov. 1948. Beirut; and NA 501. BB Palestine/11–1648, Cairo (Patterson) to Secretary of State, 16 Nov. 1948.

[37] AFSC, Foreign Service 1948, Palestine, 'Confidential Memorandum', John Devine to Ambassador Stanton Griffis (Cairo), 13 Dec. 1948.

Quakers' management, the daily refugee diet rose to 1,800 calories per day. The Quakers were distributing about 200 tons of food per day, six days a week.[38]

Powdered milk also began to reach at least some of the child refugee population by the end of December. There was a milk distribution centre at Bureij camp, operated by the Red Cross. A western observer, arriving at the end of one distribution session (which he described as being handled 'in an extremely haphazard fashion'), spoke of 'children holding old beer bottles and tin cans . . . clamouring at the window and passing their containers through it to be filled'. By March 1949 milk was reaching all the refugee children through twelve fixed distribution centres.[39]

In certain areas of the Strip lack of water emerged as a major problem. In Rafah, in mid-November, the local inhabitants were selling the refugees water 'for 20 piastres a liter'.[40] In mid-December at Maghazi camp, south of Bureij, the 3,000 refugees had water, 'but often at some distance and in inadequate quantity'.[41] By the end of December, the Egyptian Army had installed in all the camps 'water tanks made from former British boilers and petrol tanks' and water was trucked in and distributed for two hours each day.[42]

The refugees' lack of clothing was another general problem, one which became more acute as the fall weather changed to winter. The refugees arrived in the Strip over October–November with very little baggage and very few clothes. The Egyptian military governor of Khan Yunis, Captain Nagui Amin Sallam, at the end of December defined the shortage of clothes as more 'urgent' than food. The Egyptian officer in charge of refugee affairs in the Strip, Captain Wahid, said that the combination of poor nutrition and lack of clothes would make the refugees 'easy prey' to pneumonia and

[38] AFSC, Foreign Service 1948, Palestine, 'Memorandum', Gillespie Evans to Ambassador Stanton Griffis (Cairo), 26 Dec. 1948; and AFSC, Foreign Service 1948, Palestine, 'Report of Clarence E. and Lilly Pickett on their Visit to the Middle East', 3 Mar. 1949.

[39] Ibid.; ibid.

[40] AFSC, Foreign Service 1948, Palestine, 'Report of Visit by Dr Descoeudres to Southern Palestine, Nov. 11–12, 1948', 13 Nov. 1948, Beirut.

[41] AFSC, Foreign Service 1948, Palestine, 'Confidential Memorandum', John Devine to Ambassador Stanton Griffis (Cairo), 13 Dec. 1948.

[42] AFSC, Foreign Service 1948, Palestine, 'Memorandum', Gillespie Evans to Ambassador Stanton Griffis (Cairo), 26 Dec. 1948.

other diseases. Most of the refugees cut up blankets distributed by the relief organizations and turned them into 'coveralls . . . in which [they] . . . live and sleep'.[43]

Medical services in the Strip remained appalling through November–December 1948 and, indeed, into 1949, until the activation of AFSC-directed teams. At the end of 1948 the Red Cross representative in the Strip, Dr Pflimlin, estimated that 120 refugees were dying each night, mostly from starvation and dysentery. An Egyptian doctor, who refused to treat serious cases among the refugees, was quoted as saying: 'After all, this is the season for dying.'[44] Dr Pflimlin and a local Arab doctor were the only doctors regularly treating the refugees during November. Pflimlin, who was Swiss, was assisted by a Swiss nurse. A second nurse joined him in December. The doctors had almost no medical supplies.[45] In December, Egyptian army doctors occasionally visited the camps and dispensed treatment to the sick but there were no regular hours and no medical centres or clinics in any of the camps. An American observer reported that, for all his efforts, Dr Pflimlin '[has] not made any noticeable dent in the [refugees' medical] problem'.[46] Local doctors, of whom there were only a handful, were apparently unwilling to treat destitute refugees without payment.

There were several small hospitals in Gaza—one run by the Red Crescent, another by Dr Hargreaves of the proselytizing Church Mission Society—but these had been established with the Gaza Strip pre-1948 local population in mind, and from November 1948 only a handful of beds were set aside for serious refugee cases. The problems of hospitalization in the Gaza Strip were made more acute by the renewed IDF offensive of 22 December 1948–6 January 1949, in which dozens of local inhabitants and refugees—especially from the mortared Maghazi camp—were killed and injured. A western observer who visited the Red Crescent Hospital in Gaza on

[43] AFSC, Foreign Service 1948, Palestine, 'Memorandum', Gillespie Evans to Ambassador Stanton Griffis (Cairo), 26 Dec. 1948.

[44] NA 501. BB Palestine/11–1648, Cairo (Patterson) to Secretary of State, 16 Nov. 1948, and 501. BB Palestine/12–748, Cairo (Patterson) to Secretary of State, 7 Dec. 1948.

[45] AFSC, Foreign Service 1948, Palestine, 'Report of Visit by Dr Descoeudres to Southern Palestine, Nov. 11–12, 1948'. 13 Nov. 1948, Beirut.

[46] AFSC, Foreign Service 1948, Palestine, 'Confidential Memorandum', John Devine to Ambassador Stanton Griffis (Cairo), 13 Dec. 1948.

23 December reported that it was 'completely unheated and unlighted' and that the injured from the mortar attack overflowed on to the hospital porch. The hospital was staffed by two Arab doctors and a handful of 'completely untrained interns'. There was a shortage of 'everything', according to the hospital director.[47]

But, despite the almost complete absence of medical care or supplies and the onset of winter, the refugees' health situation was not as bad as western observers expected, in large measure because of what these observers defined as the refugees' natural 'hardiness'. Indeed, after the initial death toll of November 1948, it was not clear that infant, child, or adult mortality rates among the Gaza refugees from the beginning of 1949 were any higher than among the indigenous Gaza population. Some of the refugees were inoculated by the Egyptian Army against smallpox and typhoid, and no epidemics broke out during the first crucial winter months of November and December 1948 or during the first months of 1949. By February–March 1949, the refugee population's health was adequately assured by the Quaker-run teams of doctors and para-medics.[48]

Although the refugee concentrations in Transjordan and Trans-jordanian-held parts of Palestine had not been substantially increased by further fighting after the summer of 1948, conditions there too had deteriorated, mainly because of the advent of winter. The commander of the Arab Legion, General Glubb, in December wrote from Amman to J. Baker-White, MP, that 'many died in the night', after the first rain of the season. Down the road from his house eleven refugees were living in a quarry. Glubb had sent them blankets 'but their little boy of about four . . . during the night died all the same'.

The Transjordanian authorities, in concert with the UN refugee officials, had during the autumn moved tens of thousands of refugees from the Ramallah–Hebron hill-country, where snow occasionally fell in winter, to the warmer Jordan valley; about 30,000 were living

[47] AFSC, Foreign Service 1948, Palestine, 'Memorandum', Gillespie Evans to Ambassador Stanton Griffis (Cairo), 26 Dec. 1948.

[48] AFSC, Foreign Service 1948, Palestine, 'Report of a Visit by Dr Descoeudres to Southern Palestine, Nov. 11–12, 1948', 13 Nov. 1948, Beirut; and AFSC, Foreign Service 1949, Palestine, 'Report of Clarence E. and Lilly Pickett on their Visit to the Middle East', 3 Mar. 1949.

in refugee camps (and another 5,000–6,000 elsewhere) around Jericho by December.

Glubb complained at the start of December that the Jericho area refugees were getting 'no food'. This was inaccurate. According to British officials on the spot, the Transjordanian Government was distributing flour sufficient for 225 grams of bread per refugee per day, and food was arriving from the UN relief agency. More than 1,000 half-pints of milk were being prepared from powder and distributed daily for infants and nursing mothers. Two kilograms of potatoes and chunks of cheese had been distributed to each person, and a shipment of olive oil, tinned meat, and margarine had been received and was to be distributed. A shipment of 1,000 blankets had also arrived and largely had been distributed.

The Jericho camps, according to British officials, were being run efficiently by a Belgian Red Cross official, Dr Depage, but suffered from a basic lack of funds to cover such expenses as 'milling of wheat, sanitation, water supply, warehousing . . . and transport'. A Danish field hospital unit—with staff—was due to arrive, but meanwhile Depage suffered from lack of medical personnel. There were Arab doctors in Jericho but 'they were not willing to work without some form of payment' and Depage had no money.

Depage had probably created the first major experiment in organized camp life for the refugees in Palestine, a model camp. An attempt had been made to place people from the same villages together, in blocks. Arabs were given various jobs in the camp administration and cleanliness was enforced. And it appeared to work, as far as it went. But in January 1949, lack of supplies from the Disaster (Refugee) Relief Project, which had employed Depage, administrative changes, including Depage's removal, and unemployment resulted in major violence. Discontented refugees ransacked Jericho's Jordan Hotel, killed an Arab Legion officer, and wounded several people. The refugees had heard that relief money was being used to throw a party at the hotel. The camp was cordoned off by Legion armoured cars, searched, and placed under curfew. Dow, the British consul-general in Jerusalem, pressed for funds to employ and pay at least some of the camps' male population, 'paid employment [being] a most important psychological factor in maintaining the morale of male refugees and preventing them from lapsing into anarchy and Communism'. The Red Cross, which had taken over

the camps from the UN organization, apparently peferred bringing
in its own staff to using local workers.[49]

The situation in the Nablus area, controlled by the Iraqi military,
remained unchanged. 'To turn to Nablus from Jericho is to sink
from Day No. 1 of the creation back into chaos', wrote Dow. The
refugees in the Nablus–Tulkarm area refused to move to warmer
Jericho. The consul-general believed that they were being encouraged
in this course by local officials and merchants, who stood to gain by
their continued presence (charging those with money exorbitant
prices and siphoning off some of the relief money destined for the
refugees). The local authorities maintained that there were 128,000
refugees in their area; Dow thought the figure inflated, and implied
that the inflated figure stemmed from a desire to receive more relief.
At least three refugees were dying each day in the immediate
vicinity of Nablus, according to the mayor. There was only haphazard local distribution of food and milk and there were no tents.
'The Iraqi army with . . . 1 line of [Communications] of 1,500 miles
are unwilling and probably unable to do anything about them.
There is flour in the Bazaar but the refugees cannot afford it.'
The municipal authorities wanted the Transjordanian Government
to administratively take over the Nablus area but the Iraqis,
'for reasons of prestige', refused to relinquish any authority. The
consul-general recommended that a refugee camp be set up east of
Nablus and that relief supplies only be given to those refugees who
agreed to move to the camp.[50]

But the whole concept of the refugee camps (whether run directly
by a UN agency or by the Red Cross), almost from the start was
understood to be politically dangerous and insufficient. Cilento,
who more or less supervised the establishment of the camps, told a

[49] PRO, FO 371–68633 (E16038/10748/31), Glubb Pasha (Amman) to J. Baker-
White, MP (London), 2 Dec. 1948; FO 371–68683 (E16264/10748/31),
R. A. Beaumont (Consulate-General, Jerusalem,) to B. A. A. Burrows, Foreign
Office, undated, *c.* 15 Dec. 1948; FO 371–68683 (E16344/10748/31), Sir H. Dow
(Jerusalem) to Foreign Office, 21 Dec. 1948; FO 371–75417, Sir H. Dow (Jerusalem)
to Foreign Office, 16 Jan., 1949; and NA 501. BB Palestine/12–2748, 'Minutes of
Fifth Meeting with Voluntary Agencies.' Beirut, 3 Dec. 1948. Avi Plascov, in *The
Palestinian Refugees in Jordan, 1948–1957* (Frank Cass, London, 1981), 66–7,
briefly describes the switch from tents to huts—in fact, to relative permanency—of
the Jericho area camps over 1949–50.
[50] PRO, FO 371–68683 (E16344/10748/31), Sir H. Dow (Jerusalem) to Foreign
Office, 21 Dec. 1948.

gathering of representatives of voluntary relief organizations in Beirut in early December that the 'camps are not a good idea if their establishment leads to permanency or to an intensification of the feeling of isolation and frustration, which is the tendency of many of the refugees; nor are they good if they remain merely areas for the distribution of goods without any purposive trend towards a final solution'. The outbreak in Jericho in January 1949 had illustrated the danger the camps constituted to law and order, and perhaps to government, as well, in the Hashemite kingdom. A permanent solution, all understood, was possible in one of two ways: Repatriation to Israel or resettlement in the Arab states. A third possibility, of combining the two, was in fact a variation on the idea of resettlement in the Arab countries as, if Israel refused to accept and adopt the principle of repatriation (as was the case), it would certainly not accept more than a very small number of repatriates as part of any political deal with the Arabs.[51]

'THE REFUGEE THREAT' AND THE POSSIBILITY OF ORGANIZED RESETTLEMENT IN THE ARAB STATES

By the start of 1949, most US diplomats in the Middle East understood that repatriation was out of the question as Israel, simply, refused to take back the refugees. And some, such as James McDonald, had come around to the Israeli view that the alternative, of organized resettlement in the Arab countries, was indeed preferable. As William Burdett Jr., the new US consul-general in Jerusalem (who was generally unfriendly to Israel), put it: 'Political-security best served by settling refugees in Arab states. Return refugees [to Israel] would create continuing minority problem and form constant temptation both to uprisings and intervention by neighbouring Arab states.' It was the height of the Cold War and Burdett, like most other American officials, was keenly aware of the revolutionary, 'pro-Communist' potential of the hundreds of thousands of disgruntled, ill-fed, ill-housed, and stateless refugees: 'USSR may capitalize on opportunity', he wrote. Unlike most of his peers, Burdett thought that the organized resettlement of the refugees could focus on Transjordan and the Arab occupied parts of Palestine (Samaria, the

[51] NA 501. BB Palestine/12–2748, 'Minutes of the Fifth Meeting with Voluntary Agencies,' Beirut, 3 Dec. 1948.

Hebron Hills, and the Gaza district). His thinking was based primarily on his knowledge or intuition that the other Arab states—principally Syria and Iraq—would not agree to absorb any more refugees than they already had. But permanent absorption in Transjordan and the Arab-held parts of Palestine, he argued, would necessitate large-scale development schemes, including irrigation projects in the Jordan Valley, road construction, fisheries, phosphate production, and potash plants, which would provide employment. All this could be done only if the Western states poured in a lot of money.[52]

But Israel and most American officials preferred to focus on Iraq and Syria rather than Transjordan and Palestine, as the best sites for the resettlement of the refugees. The head of the US Legation in Jidda, Saudi Arabia, wrote that while, under Arab League pressure, the Saudis might accept 'a token number of Arab refugees', the kingdom would refuse to take any substantial number. There were both economic and political reasons, the latter weighing the more heavily. Saudi agriculture could absorb none and the oil industries could absorb only a handful. But more important, 'the entry of a significant number of Palestinians . . . could not help but have far-reaching political implications . . . [because] the Palestinians, under British tutelage, grew accustomed to at least a semi-free press and other beginnings of democratic institutions. They would doubtless find it most difficult to accept the situation in Saudi Arabia where the rule is absolute . . .' Hence, the senior American diplomat felt that it was neither in the Saudis' nor America's interest to press the king to take in more than a token number of refugees. As repatriation to Israel was to be ruled out, resettlement in Arab countries, 'principally Iraq and possibly Syria', should be planned. The US would have to foot the bulk of the bill, he felt. In the absence of an

---

[52] NA 501. BB Palestine/2-949, Burdett (Jerusalem) to Secretary of State, 9 Feb. 1949. Already in Aug. 1948 US diplomats were warning of the 'Communist' revolutionary potential of the Palestinian refugee communities. On 24 Aug., Keeley, the US Minister in Damascus, after a tour of camps in Syria wrote that the refugees were unquestionably 'ripe for Communist indoctrination'. See NA Record group 84, Jerusalem Consulate General, Classified Records 1948, 800—Syria, Keeley (Damascus) to Secretary of State, 24 Aug. 1948. See also NA 501. BB Palestine/3-2849, Beirut (Mark Ethridge, US representative on the Palestine Conciliation Commission) to Secretary of State (and, at his discretion, to the President), 29 Mar. 1949, for Ethridge's assessment of the refugees as a threat to the stability of the Arab host regimes.

organized resettlement scheme, the refugees would 'remain a canker in the Near Eastern states and [a threat to] American relations with them for some time to come'.[53]

Similar resistance to the permanent absorption of Palestinian refugees was expressed by Lebanon, which by the start of 1949 had 90,000–120,000 on its soil. Economically, they were an 'unbearable burden'; Lebanon suffered from 5 per cent unemployment and could offer almost no employment to the Palestinians. The US Beirut Legation concurred: 'It is difficult to see how present economic conditions in Lebanon will permit any significant absorption of refugees.' But even more important was the political aspect. 'The absorption of an alien population amounting to as much as 10 per cent of the native population . . . would create a Moslem majority and turn the entire political complexion of the country.' The Beirut Legation also pointed to Iraq and Syria, after appropriate agricultural investment and development, as the best sites for the refugees' relocation.[54]

A similar focus on Syria and Iraq for the refugees' permanent resettlement was to be found in Israeli planning at this time for the refugees' future. The proposal of the 'Transfer Committee' (set up by the government to smooth the path of the Palestinians' permanent relocation outside Israel), submitted to the Israeli Government at the end of November 1948, contained long sections detailing the absorptive capacity of the two countries. For months thereafter, committee members Ezra Danin, a senior Foreign Ministry official, and Yosef Weitz, director of the Jewish National Fund's Lands Department, promoted schemes for the absorption of the refugees in Syria, Iraq, and, to a lesser extent, Transjordan and for their employment by foreign companies with interests in the Middle East, primarily the oil corporations.[55]

The problem, however, was that Syria, with some 80,000 refugees, and Iraq, with 4,000–5,000, wanted no more. Those already in Iraq could be employed in road construction, if funds were found for such projects, wrote the American Embassy in Baghdad; their

[53] NA 501. BB Palestine/1–2149, American Legation, Jidda, to Secretary of State, 21 Jan. 1949.
[54] NA 501. BB Palestine/2–449, American Legation, Beirut, to Secretary of State, 4 Feb. 1949.
[55] For details see Essay 4 in the present volume.

continued presence would have only a 'negligible effect' on the country. But Iraq's policy was that even these 'should be returned to Palestine'. As to additional refugees, economically—with appropriate international investment in agricultural development schemes —it was 'conceivable'. But socially and politically, it would be extremely problematic. 'Despite their common Arab culture, Palestinians would be regarded as foreigners and an influx of them would not be welcome either in the towns or in the agricultural areas of Iraq. They would be likely to form another unassimilated, discontented minority group', wrote the American Embassy. The Embassy, in short, reflecting Iraqi Government opinion, thought that 'the presence in Iraq of many thousands of victims of the clash between Arabs and Jews would be certain to keep the Palestine issue burning alive for generations. In the interest of long-range security in the Middle East, the Embassy recommends therefore that every effort be made to ensure . . . that refugees who so desire be returned to their homes [in Palestine] . . . '. The Iraqis, then, wanted no more refugees, for economic and social-political reasons. Neither did the Syrians.[56]

But it went deeper than that, and this applied to Iraq as well as all or most of the Arab states, as the American representative to the Palestine Conciliation Committee, Mark Ethridge, understood. A major reason for the Arab unwillingness to properly absorb and resettle the refugees at this time was rooted in their struggle against Israel rather than in internal political or economic and social considerations. The Arabs saw in the '700,000 or 800,000' refugees a 'political weapon against the Jews. They feel they can summon world opinion [against Israel] even if some refugees die in the meantime.' The Arabs sought the refugees' repatriation, both as a demand of justice and as a means to subvert the Jewish state. Hence, it was better to leave the refugees for the time being in squalor and living impermanently than to support and implement resettlement schemes in Arab countries which would neutralize, and perhaps ultimately solve, the problem. Moreover, the refugees camped along Israel's borders (rather than in faraway Iraq), the Arab leaders told Ethridge, would provide the 'core of [an] irredentist movement that will plague all Arab states and provide basis for con-

[56] NA 501. BB Palestine/2–749, Baghdad to Secretary of State, 7 Feb. 1949.

tinual agitation to the point that there will be no possibility of having anything more than [Arab–Israeli] armistice in the Middle East'.[57]

So it was that the Palestinian Arab refugees, numbering between 600,000 and 760,000, remained *in situ*. Unwanted as repatriates by Israel and unwanted as new immigrants by the larger Arab countries, the refugees remained by and large where they had initially come to semi-permanent rest—mainly in and near the main Arab towns of Samaria and the Hebron Hills, Jericho and Transjordan, and the Gaza District, and, in smaller numbers, in Lebanon and Syria.

Through 1948 the refugees suffered; hundreds, perhaps thousands, died. Their initial absorption by the Arab states and by the eastern Palestinian and Gazan communities, and the Iraqi, Transjordanian, and Egyptian armies that controlled these areas, was anarchic and thoughtless. As most of the refugees reached these areas in the spring and summer of 1948, the need for reasonable, rain-proof accommodation was not immediately apparent. Life in olive groves, caves, private gardens, and unfinished buildings and garages was possible, given the temperate climate. Moreover, for months most of the refugees and the Arab local and national leaders continued to believe that the refugees' lot was temporary, and that they would soon be returning to their homes. Few were able to think of non-return as a realistic prospect, or to suit their thinking and actions to what was and would be needed if the Palestinians' refugeedom was to become permanent. Lastly, the Arab states—poor, corrupt, and highly disorganized, by Western standards—throughout 1948 were involved in a major war with Israel, their main energies and thinking focusing on the war effort rather than on the alleviation of the refugees' misery. The refugees were a side-issue, and were regarded mainly as an added burden for the Arab armies and states. The alleviation of their physical plight was left largely in the hands of local Arab authorities and voluntary committees, though in Transjordan, Transjordanian-occupied territory, and Lebanon, the state provided a basic quantity of food for daily sustenance, but little else. There was unhappiness about the influx of refugees in various localities; they brought filth, disease, and demoralization.

---

[57] NA 501. BB Palestine/3–2849, Beirut (Ethridge) to Secretary of State, 28 Mar. 1949.

Why should Egyptian and Transjordanian boys fight for the rights
of Palestinians who had fled without a fight? some argued. Good
ladies organized to give aid, blankets, food, money. But local
merchants and, apparently, officials exploited the refugees, hiking
prices and siphoning off a proportion of the international and local
relief contributions.

Attempts by Arab governments, such as Syria and in some areas,
Transjordan, to get the refugees to go back were not pressed home.
And, in any case, Israel was firmly resolved, from June 1948, not to
allow the refugees to return, either in dribbles or *en masse*. Attempts
by refugees to return were repulsed by fire. Moreover, many of the
refugees, during the summer and autumn of 1948, preferred to stay
put for some of the same reasons that had prompted them to leave
Israeli-controlled areas in the first place—fear of the Jews and
unwillingness to live as a minority in a Jewish state. Some, at least
for a time, expected to return to their homes in the wake of
conquering Arab armies.

International relief efforts on behalf of the refugees were slow off
the mark, partly also because of a belief that the refugees would
soon be repatriated. But as winter approached and everyone grew
to accept that repatriation was not on the cards, tent camps were set
up to accommodate the refugees. From the first, these were regarded
—by the international relief bodies and the Arab host countries—
as a temporary arrangement; the danger of the presence of hundreds
of thousands of disgruntled, poorly fed and poorly housed refugees
to the Arab host regimes was clear to all. By the winter, and only
just, the UN and the other international agencies had overcome the
bureaucratic muddle and the bottlenecks in Beirut and elsewhere,
and a minimal if sufficient amount of relief aid began reaching the
refugees. The UN managed to put its act together with the belated
formation of the UNRPR, and the Arab states, armies, and municipalities during the winter of 1948-9 were almost completely
relieved of the burden of care for the basic physical necessities of the
refugee communities.

But this was seen by the international community as merely a
temporary situation, which required a major political initiative and
solution. Pressures on Israel to allow back all or a large number of
the refugees increased over the winter and spring; but, given its
victories over the Arabs, so did the new state's strength, and its

ability to resist such pressures successfully. Repatriation was just not on. But neither, it turned out, was its alternative, organized resettlement in the larger (and largely under-populated) Arab countries— the solution by the start of 1949 favoured by Israel and most senior American and British officials (who also wanted to see Israel take back some refugees, if only as a political gesture of conciliation towards the Arabs). In the course of the summer and autumn of 1949, Iraq, Syria, and the other major Arab countries resisted the dangled bait of large economic aid linked to readiness to absorb large numbers of refugees. They had good economic, social-demographic, and internal political reasons, the latter—the threat of revolutionary destabilization—apparently predominating. But also, as Ethridge perceived, there was a need born of the context of conflict with Israel—to use the refugees as a weapon against the Jewish state. The hundreds of thousands of refugees, if left in misery, would continue to demand repatriation, and would find support for this in the international community. If allowed back, they could destabilize the Jewish state. If not allowed back, their existence would eat at world support for Israel and would ignite an endless and violent irredenta, which would leave the Arabs and Israel in permanent conflict. Ethridge pressed for American pressure on Israel to take back some 250,000 of the refugees to neutralize this irredentist-political threat and to do at least partial justice to the Palestinians. But Israel, having won the war and got rid of its potentially destabilizing, large Arab minority was in no mood for compromise on the matter, a compromise seen by almost all of Israel's security-minded leaders as in itself a mortal threat to the country's well-being. The refugees remained in the camps.

9

The Transfer of Al Majdal's Remaining Arabs to Gaza, 1950

The Arab town of Majdal, site of the second millenium CE Philistine port of Ashkelon or Ascalon on the Mediterranean coast between Jaffa and Gaza, had a population of about 10,000 at the start of the 1948 war. During Operation Yoav, the IDF repeatedly shelled and bombed the town and, on 4 November, conquered it. Majdal's Egyptian garrison, and much of the population, had fled the town already on 30 October.

The occupying Israeli force was ordered by OC Southern Command, General Yigal Allon, to expel the town's remaining Arabs but the local commanders apparently baulked, and the order was eventually rescinded.[1] Something over 1,000 remained, their number growing to 2,400–2,700 in the following two years. The increase was due to births, infiltration of refugees back into the town from the Gaza Strip, and the movement to the town of remainder communities of Arabs from elsewhere in the South of Israel.

Throughout the 1948–9 period, Allon 'demanded . . . that the town be emptied of its Arabs'. New Jewish immigrants began moving into Majdal—once again called Ashkelon—in 1949 and were settled in abandoned Arab houses. The immigration absorption and settlement authorities wanted as many empty Arab houses as possible. The IDF did not want an Arab-populated town near its frontier with Egypt. Sometime early in 1950, the IDF General Staff formally demanded government authorization for the eviction of Ashkelon's Arabs.[2]

---

[1] Notes written by Aharon Cohen during the meeting of Mapam's Political Committee, 11 Nov. 1948, in HHA-ACP 10.95.10 (6); and 'About the Eviction/ Evacuation [pinui] of Migdal-Gad's [that is, Majdal's] Arab Inhabitants', 22 Oct. 1950, an unsigned two-page memorandum, probably by officials of the Israel Foreign Ministry's Research Department, in ISA, FM 2436/5bet.
[2] 'About the Eviction . . . ', 22 Oct. 1950, in ISA, FM 2436/5bet.

The interdepartmental government Committee for Transferring Arabs already in February 1949 had decided in principle on the need to clear Majdal of Arabs, but desisted for political reasons (Israel at the time was seeking admission to the United Nations) from pressing the matter.[3]

The General Staff's demand for the transfer of Ashkelon's Arabs was based on (*a*) a desire to end the ceaseless and troublesome infiltration back and forth across the Egyptian–Israeli border; (*b*) Ashkelon's position as a 'source of information and refuge for Arab infiltrators' and as a way-station for infiltrators bound for Ramle, Lydda, and Jaffa; (*c*) the desire of Ashkelon's Arabs to see 'the return of Arab rule in their city' (hinting at their potential as a Fifth Column); (*d*) the cost of maintaining an efficient military government in the town to oversee the Arab population and of paying for the upkeep of the unemployed and the social cases in the Arab community; and (*e*) the assessment that the continued occupancy of the town by Arabs necessarily 'politically and militarily weakened [Israel's] hold over the town and its environs'.[4]

In early 1950 the authorities—apparently in a formal Cabinet decision—approved the 'orderly' transfer of Majdal's remaining Arabs to sites inside Israel, with the largely abandoned former Arab towns of Ramle and Jaffa being earmarked for their absorption.[5]

During the spring of 1950 Ashkelon's Arabs were repeatedly informed by the Israeli military governor, Lieutenant-Colonel Yehoshua Varbin, and his subordinates that they would shortly have to evacuate the town. As the investigating UN officer put it in a report on 4 September 1950: 'They have been warned two or three times that they all would have to leave El Majdal within a [*sic*] near

---

[3] 'Protocol of the Ninth Meeting of the Committee for Transferring [Arabs]', 25 Feb. 1949, in ISA, MAM 1322/22; and CZA A246-14, pp. 2584–5, entry for 23 Feb. 1949. The Committee for Transferring Arabs (sometimes also referred to in the documentation as the Committee for Transferring Arabs from Place to Place) was set up at the end of 1948 as an interdepartmental body consisting of representatives of the Interior Ministry, the Agriculture Ministry, the police, the Defence Ministry, and the Jewish National Fund (Yosef Weitz). Its task was to resettle Arabs who had been displaced by the hostilities but still lived in Israeli territory in new sites. In the course of 1949–50, the committee resettled 'remainders' from such Galilee villages as Khisas, Ja'una, and Qeitiya in Akbara; from Al Bassa and Kabri in Mazra'a; from Kafr 'Inan and Farradiya in Tur'an; from Saffuriya in Ar Reina, and so on.

[4] 'About the Eviction . . .', 22 Oct. 1950, in ISA, FM 2436/5bet.

[5] Ibid.

future.' An initial convoy of Majdal residents—some 200 souls originally from the village of Qatara displaced to Majdal during the 1948 war—were transferred in June or July to Ramle and resettled there in abandoned Arab houses.[6]

But at some point during that spring, General Moshe Dayan, Allon's successor as OC Southern Command, and his subordinates decided to direct the evacuation of Majdal's Arabs out of Israel, toward the Gaza Strip, rather than to sites inside the country. Dayan received authorization for this from Prime Minister and Defence Minister Ben-Gurion at a meeting between the two men on 19 June—overcoming objections to the transfer scheme by Histadrut Secretary General, Pinhas Lavon. According to Dayan's memoirs, Ben-Gurion during their talk had set one condition, 'one which I, too, had in mind— . . . that we would make the transfer only if the Majdal Arabs themselves agreed to it'.[7] The exodus to Gaza, in fact, began already on 14 June.

[6] Major V. H. Loriaux, acting Chairman of the Egypt–Israel MAC, to UNTSO Chief of Staff, 4 Sept. 1950, in ISA, FM 2436/5bet; General Moshe Dayan to IDF Chief of General Staff, early Oct. 1950, in ISA, FM 2436/5bet. Dayan wrote that the 'military governor's semi-official announcements concerning the evacuation of Majdal by all its Arab inhabitants' was among the 'precipitating factors' (*gormim medarbenim*) in the eventual evacuation.

[7] Aluf Ben, 'Halchu Le'Aza', in *Ha'ir*, 23 Sept. 1988; and Moshe Dayan, *Story of My Life* (London, edn., 1977), 160–1. Dayan in his memoirs, presents the Majdal transfer as a joint initiative of the Majdal Arab leaders and himself. 'After talking to them, I found that the majority wished to move to the Gaza Strip . . . while some wanted to go to other towns inside Israel. I thought this an admirable solution.' Dayan relates that he then approached the Egyptians, who 'readily agreed . . . I promptly presented the proposal to the [IDF] General Staff. They were sluggish in their response, so I went to see Ben-Gurion, and he gave his approval.' But just as the final preparations for the transfer were getting under way 'I [Dayan] received an urgent telephone call from the deputy chief of staff telling me to suspend activities . . . It turned out that . . . Lavon . . . had demanded that the defence minister scrap the plan. The Histadrut [the general trades-union federation] had its own proposal, which offered the promise of employment for the Majdal textile workers but not to the others. My proposal [to transfer them to Gaza] would provide the opportunity of immediate employment for all the workers of Majdal and would also meet the social and cultural needs of the entire community. Lavon and I were then summoned to Ben-Gurion, who heard both sides and decided in my favour.' Nowhere does Dayan explain why the IDF General Staff reacted sluggishly to the OC Southern Command's proposal or why Lavon rejected it—though presumably both were wary of the possible reaction of the Western governments and international public opinion to the proposed transfer. In any case, for various reasons, I tend to doubt the accuracy or comprehensiveness of this explanation of the decision-making process in the transfer, though there can be no doubt that Dayan played a major role in initiating the process.

A web of interlocking 'internal' / objective and 'external' / Israeli-directed pressures and inducements prompted and fuelled the emigration of Majdal's Arabs to Gaza, the repeated threats of imminent transfer by the town's military administration being, perhaps, only one of the more potent ones.

Israeli observers at the time believed that the transfer of the 200 to Ramle had had a catalytic effect on the Arabs of Majdal, 'triggering among the remaining inhabitants of the city a growing tendency to leave . . .'[8]—much as, during the 1948 exodus, the flight of one part of a village's or town's population triggered the flight of the rest of the population.

But there were more basic, cogent factors at play. Among the 'internal'/objective reasons were the Majdal Arabs' feeling of being cut off from their family members and friends already living in the Gaza Strip (most of whom had fled the town two years before) and, in a larger sense, being cut off from the wider Arab world. Two years as an abject, isolated minority community in Israel had induced in Majdal's Arabs a collective depression and purposelessness. Dayan put it this way:

Much time has passed and the chances of a quick [Israeli–Arab peace] settlement have weakened. Most of the [Majdal] families have members in Gaza and even in Egypt . . . No less than the separation of the families, there pressed upon the Arabs of Majdal their lowly and debased social condition. From free men and masters among their people they have turned into a miserable community living in a lifeless town. With all the difficulty of life in the Gaza Strip, the social and national feeling of the Arabs living there is better than that of the Arabs in Majdal.[9]

The objective factor of dismemberment of families due to the war was exacerbated by the rigid guidelines Israel applied in its at the time much vaunted 'Family Reunion Scheme'. The scheme, a sop designed to parry Western demands for Palestinian refugee repatriation, provided for the return to Israeli territory of 'family' members on humanitarian grounds. 'The bulk of Majdal's population consisted of fragments of families' But 'Israel interpreted the concept of "family", in this context, in the narrow European sense [that is,

---

[8] 'About the Eviction . . .', 22 Oct. 1950, in ISA, FM 2436/5bet. This point was also made by Yehoshafat Harkabi in his notes, appended to Harkabi to Ziama Divon, 24 Sept. 1950, in ISA, FM 2436/5bet.

[9] Dayan to Chief of General Staff, early Oct. 1950, in ISA, FM 2436/5bet.

children, wife, and parents] and not according to the wide Arab–oriental definition which included in the family brothers, sisters, relatives, uncles, etc.', explained a Foreign Ministry report.[10] So Majdal's Arabs could not expect to be reunited with most of their family members in Majdal—leaving only the option of (exodus and) reunification in the Gaza Strip.

There were also good, objective reasons for the departure. In contrast with the prevailing unemployment and the restrictions placed by the Israeli authorities on employment in Majdal, there was a 'flowering of the weaving industry' in the Gaza Strip. Many Majdalites were weavers and they expected to find lucrative employment in the Strip. The Israeli authorities pointedly announced that departees would be allowed to take their looms with them.[11]

During the spring and summer of 1950, the Arabs of Majdal also feared the imminent eruption of a 'second round' of Arab–Israeli hostilities and that, living in a town near the Israeli–Egyptian frontier, they would be caught in the crossfire and, perhaps worse, might be considered 'traitors' by their possibly victorious Arab brothers. Some Israeli sources even speak of the Majdal Arabs' fears, against the backdrop of the outbreak, in June, of the Korean War, of being harmed in a Middle Eastern offshoot of a third world war.[12] Egyptian propaganda during 1950 frequently spoke of a coming 'second round' and it is likely that in the early months of 1950, belief among the Majdal Arabs in the imminence of such a 'second round' was promoted by the local Israeli authorities. This was probably one of the major components of the 'black propaganda' campaign used by the authorities to fuel the exodus. As one Foreign Ministry official put it at the time, 'we helped [foment] the panic and the general [depressed] feelings by [use of] black propaganda [*ta'amulat lahash*]'[13]—a method, incidentally, utilized successfully

[10] 'About the Eviction . . .', 22 Oct. 1950, in ISA, FM 2436/5bet.

[11] Ibid.; and Dayan to Chief of General Staff, early Oct. 1950, in ISA, FM 2436/5bet.

[12] 'About the Eviction . . .', 22 Oct. 1950, in ISA, FM 2436/5bet;Harkabi to Divon, 24 Sept. 1950, in ISA, FM 2436/5bet; and S. Sverdlov 'Here are the Numbers of Arabs who Moved from Migdal-Gad to Gaza and Transjordan', a one-page memorandum, 24 Sept. 1950.

[13] US Department, Israel Foreign Ministry, to Israel Embassy, Washington, DC, etc., 17 Sept. 1950, in ISA, FM 2474/13aleph; and, for Egyptian 'second-round' propaganda, see 'About the Eviction . . .', 22 Oct. 1950, in ISA, FM 2436/5bet.

by the Haganah to promote flight by a number of Arab communities in Palestine, particularly in eastern Galilee, during the 1948 war.

Dayan, in October 1950, informed his superior, General Yadin, the IDF Chief of General Staff, that he 'had heard . . . no complaints from the [Majdal] Arabs about the Israeli authorities and their day-to-day treatment [of the Arabs]. Indeed, I found that they had great personal faith in the military governor and his staff . . .' But clearly a variety of powerful pressures and inducements were brought to bear on Majdal's Arabs to persuade them to emigrate. Dayan at the time put it this way:

There was no easing by Israel of restrictions . . . [on their freedom of] movement, on their employment in areas far away from the town and in the restoration of their property, [and this] led the Arabs to the conclusion that in the existing circumstances (so long as there was no peace) there was no chance of the lifting of the restrictions and of a change in their treatment by the Israeli authorities.[14]

'It is not right [to say that] the Arabs of Majdal were expelled . . .', wrote a Foreign Ministry official at the time. True, Israel had used black propaganda to encourage the departure. 'But no other measures [to induce flight] were taken. No Arabs were ejected from Majdal by force . . .'.[15]

But this description was somewhat misleading. Major V. H. Loriaux, the UN truce observer and sometime acting chairman of the Egypt–Israel Mixed Armistice Commission, interviewed some of the evacuees shortly after they reached the Gaza Strip. He was told that the Majdal Arabs, soon after being warned that they would shortly have to leave, were charged '1,650 Israeli pound [*sic*] for drinkable water (it was free of charge previously)', and their 'rations distributions were often delayed'. Moreover, their freedom of movement was severely restricted and they were penned in in what amounted to an Arab ghetto, with checkpoints and barbed-wire barriers. They had little or no contact with the surrounding, growing Jewish population.[16]

In a second letter, based on further interviews with Arabs transferred to Gaza, Loriaux said that Ashkelon Arab orange-grove

[14] Dayan to Chief of General Staff, early Oct. 1950, in ISA, FM 2436/5bet.
[15] US Department, Israel Foreign Ministry, to Israel Embassy, Washington, DC, 17 Sept. 1950, in ISA, FM 2474/13aleph.
[16] V. H. Loriaux to Chief of Staff, UNTSO, 4 Sept. 1950, in ISA, FM 2436/5bet.

owners were not allowed by the Israelis to sell their produce (perhaps in order to bar competition with Jewish grove owners) and received no wages. 'But they were authorised [? allowed] to cultivate vegetables on an allotment in the groves and eventually [to] sell the surplus.'[17]

The most persuasive, positive inducement to departure was, apparently, the Israelis' willingness to exchange the Majdal Arabs' Israeli pounds on a one-for-one basis with British Mandate Palestine pounds. (The rate at the time in Egyptian-controlled Gaza was I£3.5 or I£4 for each Mandatory pound). The departing Majdal Arabs, who had amassed quite a bit of money during their two years of life in Israel, by 22 October had changed I£160,000–I£170,000 into Mandatory pounds at this highly favourable exchange rate, 'which turned them into rich men in the Gaza Strip and enabled them to purchase property and to deal in commerce'. They were each allowed to take out of the country 120 Mandatory pounds. Moreover, these new refugees as a matter of course knew that they would receive (and received) help from the absorbing UN agencies upon reaching the Strip, without regard for their relatively favourable financial situation.[18]

It seems that the Israeli authorities also facilitated the Majdal Arabs' sale of those possessions that they decided not to carry with

---

[17] V. H. Loriaux to Colonel Bossavy, Chairman of the Egypt–Israel MAC, 3 Oct. 1950 (in ISA FM 2436/5bet). At some point Loriaux, apparently, also complained that there had been cases of Arabs who refused to move to Gaza being jailed. Israel denied this. In a formal note to the UN officer chairing the Egypt–Israel MAC, an Israeli representative, Captain Johanan Mannheim, wrote: 'Regarding the 7 Arabs who according to the statement were put in prison for refusing to sign a declaration stating their willingness to go to Gaza, I have been informed that in fact 7 Arabs were arrested in connection with certain violations of security laws. Their arrest had no ['no' is underlined] connection with the transfer of Arabs to Gaza. This is proven by the fact that although these Arabs have long since been released, they have not been transferred to Gaza.' See Mannheim to Colonel Bossavy, 16 Oct. 1950, in ISA, FM 2436/5bet.

On the basis of Major Loriaux's reports, General Riley reported that the reasons for the Majdal Arabs' agreement to move were: 'A. Since occupation of Majdal by Israel, Arabs are kept in special quarters. B. Shopkeepers not allowed to renew stock. C. Proprietors not permitted to enter their houses, lands, or groves. D. Arab rations are inferior to Israeli rations. E. Rumours are spread amongst Arabs that Majdal will become military zone. F. Many Arabs wished to stay, but found living conditions impossible through continuous vexations.' (See Riley to A. Cordier, 13 Sept. 1950, United Nations Archive, DAG-1/2.2.5.2.0-1.)

[18] 'About the Eviction . . .', 22 Oct. 1950, in ISA, FM 2436/5bet; and US Department, Israel Foreign Ministry, to Israel Embassy, Washington, DC, 17 Sept. 1950, in ISA, FM 2474/13aleph.

them into exile. 'The possibility of selling the moveable possessions at very high prices facilitated the Arabs' departure', wrote Dayan at the time. He estimated that each family sold some 'I£50–I£80' worth of goods,

and sometimes more. The Yemenite [Jewish] inhabitants of Majdal and other places, new immigrants, are paying fantastic prices—I£10 for a used blanket, I£16 for a shabby, small cupboard, I£4 for a small chicken, etc. The reason for this is the difficulty in obtaining these goods in the marketplace and not knowing the value of money . . . With the money the Arabs received for their goods in Majdal they will be able to buy other [goods] in Gaza and also live for a long time on the difference . . .[19]

Dayan summed up the 'internal' and 'external' impulses behind the emigration / transfer thus:

I think that the evacuation of Majdal's Arabs is a natural phenomenon of a pointless situation on the one hand and convenient conditions . . . for emigration on the other. The mass impetus to go which took hold of the whole community also contributed to this. It is worth noting that the emigrants are not depressed and that their relations with the authorities and the many preparations for the transfer are characterized by a lighthearted spirit and, almost, conviviality. The Arabs understand that in the absence of a permanent peace there is no point to the present situation and they do not blame the local authorities in Majdal.[20]

During the summer, the leaders of the Israel Communist Party tried to persuade Majdal's Arabs to reconsider and to stay put. They particularly argued against a move to other sites inside Israel where, said the Communists, the Majdalites would be prey to the whims of the local Israeli authorities. According to one Israeli source, this only contributed to the Arabs' eventual decision to emigrate to Gaza.[21]

The transfer was conducted in an orderly manner. A committee of three local Arab dignitaries helped in the registration of the transferees and in the organization of the sale of their property and their transport. None of the transferees sold their lands or houses (which, like the real estate of the 1948 refugees, fell into the hands of Israel's Custodian for Absentees Property).

[19] Dayan to Chief of General Staff, early Oct. 1950, in ISA, FM 2436/5bet.
[20] Ibid.
[21] 'About the Eviction . . .', 22 Oct. 1950, in ISA, FM 2436/5bet; and David Ben-Gurion diary, entry for 24 Aug. 1950, in Ben-Gurion Archive.

The Majdal Arabs were transferred in Israeli trucks and buses in convoys that left Majdal's town hall generally once a week, starting on 14 June. In the first convoy for Gaza 38 Arabs left; 7 left that day for Jordan. Another 36 left for Gaza on 18 June; 115 went on 23 July; 165 on 8 August; 146 on 23 August; 192 on 30 August. In August, 79 Majdalites left for Jordan. That month, Ben-Gurion recorded in his diary: 'The Arabs are fleeing Majdal-Gad. 1,500 are left. They are going to Gaza and Hebron. The Egyptians are receiving them. They are being allowed to sell their possessions and to exchange money.'[22] On 4 September, 145 left for Gaza; on 11 September, 119; on 19 September, 115; on 26 September, 194; on 2 October, 201; on 9 October, 374; and on 12 October 229. About 100, preferring Israel, eventually moved to Lydda and Ramle. By 22 October, only 32 Arabs were left in Majdal. (Almost all of these moved out during the following months.)[23]

A curious feature of the transfer was the relative docility with which the Egyptians initially reacted to it. The Egyptians appear to have believed that the Majdal transferees had, in fact, moved to Gaza on their own volition. Certainly, the Egyptians that summer behaved as though they were reluctant to heat up their relations with Israel. But fairly soon after reaching the Strip, the new refugees began complaining that they had 'been compelled to leave . . . Some declared that they have been expulsed from their house on behalf of Israeli families [*sic*].'[24]

This may have prompted the Egyptians' first, hesitant raising of the matter at the meeting of the Egypt–Israel Mixed Armistice Commission (MAC) on 15 August. They complained not about 'expulsion' but about the fact that Israel was dumping these refugees on their border without prior notification or co-ordination. The Egyptian representative said: 'We complain against [moving] all these people to our side [*sic*]. You are forcing us to accept them. We

---

[22] David Ben-Gurion diary, entry for 24 Aug. 1950, in Ben-Gurion Archive.

[23] 'About the Eviction . . .', 22 Oct. 1950, in ISA, FM 2436/5bet. However, another source, Mannheim to Colonel Bossavy, 16 Oct. 1950 (in ISA, FM 2436/5bet), states that there are 'some hundreds of Arab [still] living in Majdal . . . today'. From the available evidence it is impossible to determine which claim—'32' or 'some hundreds'—is correct for mid-Oct. 1950. What is certain is that by the same time the following year, no more than a handful of Arab or half-Arab families were left in the town.

[24] V. H. Loriaux to Chief of Staff, UNTSO, 4 Sept. 1950, in ISA, FM 2436/5bet.

cannot refuse to take them, because we are human. All what we want is a previous notice from you when you send them, so that we can be ready to meet them.'[25] Thenceforth, Israel gave the Egyptians advance notice of each convoy, and Egyptian officials and trucks were ready to meet the transferees at the frontier crossing at Kilometre 95. One Israeli source later wrote that 'the fact that the Egyptian army came to receive them at the border, served for those still remaining [in Majdal] as further confirmation that the decision of the departees was justified'.[26]

But the matter was not to end there. Fuelled by reports from his observers (such as Loriaux), the chief of staff of the UN Truce Supervision Organization, Lieutenant-General W. B. Riley, USMC, on 21 September issued an unusual public condemnation of Israel's ongoing 'expulsion' of the Majdal Arabs. He added that his officers' investigation had revealed 'that the Arabs alleged they had no desire to go to Gaza, but were required to sign a statement agreeing to go to Gaza, never to return to Israel and abandoning all property rights'. Riley was apparently prompted to take this stand on the Majdal Arab transfer by Israel's simultaneous expulsion, on 2 September, of hundreds ('4,000', according to the UN) of members of the Azazme beduin tribe from the Negev into Egyptian Sinai.[27]

Israel reacted by denying that the Majdal Arabs had been expelled; 'no pressure has been brought on any Arab' to leave Majdal, Israel declared. (As to the Azazme, Israel argued that they were troublesome 'infiltrees', not legal residents, and therefore Israel had every right to deport them.)[28]

Riley's position and generally deteriorating Israeli–Egyptian relations soon prompted the Egyptians to formally lodge complaints about the 'expulsion' from Majdal (and of the Azazme—and linking the two made the Israeli-argued 'voluntarism' of the Majdal transfer that much more dubious) at the UN Security Council and in the MAC. On 17 November 1950 the Security Council condemned Israel on both counts and on 30 May 1951 the Egypt–Israel MAC called on Israel to repatriate the 1950 Majdal transferees. Israel, of

[25] Harkabi to Divon, 24 Sept. 1950, in ISA, FM 2436/5bet; and M. S. Comay (Tel Aviv) to E. Elath (London), 6 Nov. 1950, in ISA, FM 2592/18.
[26] 'About the Eviction . . .', 22 Oct. 1950, in ISA, FM 2436/5bet.
[27] 'General Riley's Statement on the Expulsion of Beduin', 22 Sept. 1950, Israel Foreign Ministry memorandum, in ISA, FM 2402/12.
[28] Ibid.

course, rejected the decision and, indeed, denied the Arab–UN charges. As its representative to the MAC, Captain Johanan Mannheim put it:

the transfer of Arabs from Majdal was initiated by the Arabs themselves who were desirous of joining their families in Gaza . . . There can be no . . . justification of the word 'expelled' used in the Egyptian complaints. Not a single Arab has been forced to go to Gaza . . . There is no justification for the complaints regarding distributions of rations. The Arabs received the same rations as the Jewish inhabitants. [Israel instituted an austere, universal regime of rationing following 1948.] Any delays in distribution were equal to [*sic*] both. The complaint that the Jewish population were receiving 7 eggs per week and the Arabs 2, is completely false. The general ration is 2 eggs per week. The complaint regarding payment for water is also false. Any payments made were for arrears of taxes, and were collected by the Arab local council . . .'[29]

Many of the Majdal refugees apparently soon after reaching Gaza changed their minds and wanted back. Perhaps Egyptian officials or fellow Palestinians put them up to it. Or perhaps conditions in the Gaza Strip were persuasive enough. But the gates were now firmly barred. Israel had attained an Arab-free Ashkelon and was in no mood to accede to a return to the *status quo ante*.

Dayan stressed that there was no possibility of agreeing to the return of Arabs, even the smallest number, to the city. The matter is linked to the question of a return of property, and is intimately linked to the infiltration situation in that area [which involved a great deal of theft, smuggling, acts of robbery and the occasional murder of Jews by Arab marauders from the Gaza Strip].

Dayan thought that if the UN bodies persisted in calling for the return of the Majdal refugees, 'Israel would have to reject all the candidates for return during the joint [Israeli–UN] investigation [of each candidate] which we would have to demand'. (The MAC deision had ruled in favour of the return to Majdal of those Arabs the committee would deem 'had the right to return'. The legal adviser of the Israel Foreign Ministry, Shabtai Rosenne, suggested that this proviso 'would possibly provide a basis for instituting a

[29] Captain Johanan Mannheim to Colonel Bossavy (Bossary?), 16 Oct. 1950, in ISA, FM 2436/5bet.

preliminary investigation and [enabling] the rejection of people [that is, candidates for return] during the investigation'.)[30]

Majdal became Ashkelon, irreversibly, in 1950. The subsequent three-sided debate (Israel, Egypt, and UN) about whether the departure of the town's remaining Arabs had been 'voluntary' or coerced was something of an irrelevance and unfruitful to boot. The UN calls for a return were never implemented and the Majdal transferees were fated to linger on, for decades, indefinitely, in their new homes in Gaza's refugee camps.

What is clear is that Israel wanted this last concentration of Arabs in the southern coastal plain to leave and engineered their departure. Well-tried measures of pressure, falling short of outright expulsion, were employed, including 'black propaganda' ('God knows what will befall you if you stay'), announcements of imminent eviction, and a variety of constraints and restrictions (ghettoization enforced by sentries and barbed wire, physical isolation, control of movement, and curbs on employment). To these were added a variety of 'carrots', the most important of which was the one-time offer of very generous exchange rates for the departees' Israeli currency.

This said, the 'internal', Arab factors underlying the mass exodus should not be discounted. The remaining Majdal Arabs, an isolated and, by all accounts, miserable community, by 1950 were eager for reunification with their relatives in Gaza and, in a wider sense, for a return to life in an Arab environment and polity. The draw of what they were informed would be better employment prospects in Gaza was an added inducement to departure.

Less clear are the Israeli decision-making process leading up to the exodus and the Egyptian reponse to the exodus. The driving force behind the Israeli decision was the army's Southern Command, headed by General Moshe Dayan. It is likely that the police and the General Security Service (the Shin Bet) were also eager to see the backs of this troublesome concentration of Arabs. Dayan, in directing the exodus towards Egyptian-ruled Gaza rather than towards the Israeli town of Ramle, no doubt received the backing of Prime Minister Ben-Gurion. The only caveat—probably dictated by con-

---

[30] S. Rosenne, legal adviser Israel Foreign Ministry, to Foreign Ministry Director General, 20 May 1951, in ISA, FM 2439/8; and Lt.-Colonel Michael Hanegbi to IDF staff officer in charge of MACs, Feb. 1952, in ISA, FM 2438/6.

siderations of diplomacy and public-relations image—was that the transfer must be carried out with the Majdal Arabs' agreement. Dayan would not have needed Ben-Gurion's explicit approval of the 'black propaganda' methods employed (though he surely must have obtained Treasury agreement to the offer of the unwonted exchange rate).

But did the transfer receive the Cabinet's a priori blessing? The evidence indicates Cabinet approval of a transfer to Ramle, Lydda, or Jaffa; there is no evidence that the Cabinet subsequently formally approved the switch of destination to Gaza. At the same time, the absence of public or even internal party political debate and back-biting during and in the wake of the Majdal transfer seems to indicate that the operation in effect enjoyed a broad political consensus. Moving the Majdal Arabs to Gaza was in all likelihood perceived by most if not all the ministers as a clear, important Israeli interest, in terms of demography (less Arabs), day-to-day security (less infiltration and smuggling back and forth), strategy (no Arab concentration to serve as a way station between Gaza and Tel Aviv for potential invaders from the south), and immigration absorption and town planning (more space for Jewish immigrant settlers and one less problem to overcome in building or rebuilding Ashkelon).[31]

As to the Egyptian response to the Majdal exodus, one can only guess at the reasons for the switch from initial, passive acceptance of the transfer, and effective co-operation with Israel in the matter, to the subsequent condemnations of 'the expulsion'. These probably lay more in the sphere of general Egyptian–Israeli relations, Egyptian–UN relations, and Egyptian–Arab relations than in anything to do with the actual transfer. But only the opening of the Egyptian archives can provide an authoritative explanation.

---

[31] The protocols and minutes of Israeli Cabinet meetings are closed for at least 50 years.

SELECT BIBLIOGRAPHY

Primary sources

American Friends Service Committee Archive (AFSC), Philadelphia.
David Ben-Gurion Archives (DBG Archives), Sdeh Boqer, Israel.
Central Zionist Archives (CZA), Jersualem, Israel. Papers of the Political
Department of the Jewish Agency, protocols of the meetings of the
Jewish Agency Executive and of the Jewish National Fund Directorate,
Eliezer Granovsky Papers, manuscript notebooks of Yosef Weitz diary,
etc.
Sir Alan Cunningham Papers (CP), Middle East Centre Archives, St
Antony's College, Oxford.
Hashomer Hatza'ir Archives (HHA), Givat Haviva, Israel. Papers of the
Kibbutz Artzi, Mapam (Political Committee, Mapam Centre protocols),
Aharon Cohen Papers (HHA–ACP), Meir Ya'ari Papers, etc.
Histadrut Archive, Va'ad Hapoel Building, Tel Aviv.
Archive of Ihud Hakibbutzim Vehakvutsot, Kibbutz Hulda.
Israel State Archives (ISA), Jerusalem. State papers of the Agriculture
Ministry (AM), Foreign Ministry (FM), Justice Ministry (JM), Minority
Affairs Ministry (MAM), Prime Minister's Office (PMO).
Jabotinsky Institute (JI), Tel Aviv. Papers of the IZL, LHI, Revisionist
Movement, and Herut Party.
Jerusalem and East Mission Papers (JEM), Middle East Centre Archives,
St Antony's College, Oxford.
Kibbutz Meuhad Archives (KMA), Ef'al, Israel. Papers of the Kibbutz
Meuhad (protocols of meetings of the movement and Ahdut Ha'avodah
institutions), Aharon Zisling Papers (KMA–AZP), Palmah Papers (KMA–
PA).
Labour Archives (LA) (Histadrut), Lavon Institute, Tel Aviv.
Labour Party Archives (LPA), Beit Berl, Israel.
National Archives (NA), Washington DC, State Department Papers.
Public Record Office (PRO), London. Papers of the Cabinet Office (CAB),
Colonial Office (CO), Foreign Office (FO), and War Office (WO).
United Nations Archives (UN), New York.
Yosef Weitz Papers, at the Institute for Settlement Research, Rehovot.

Individual kibbutz archives—Mishmar Ha'emek, Ma'anit, etc.
Municipal archives in Israel—Haifa (HMA), Tiberias.

Inte· views

Yigael Yadin, Yitzkah Ben-Aharon, Yehoshua Palmon, Ya'acov Shimoni, Moshe Carmel, Eliezer Be'eri (Bauer), Binyamin Arnon, Aharon Bar-Am (Brawerman).

Published primary sources

BEGIN, MENACHEM, *Bamahteret, Ketavim* (In the Underground, Writings and Documents), 4 vols. (Tel Aviv, 1959).

BEN-GURION, DAVID, *Behilahem Yisrael* (As Israel Fought) (Tel Aviv, 1952).

—— *Medinat Yisrael Hamehudeshet* (The Resurgent State of Israel), 2 vols (Tel Aviv, 1969).

—— *Yoman Hamilhama-Tashah, 1948–1949* (DBG–YH) (The War Diary, 1948–1949), 3 vols., eds. Gershon Rivlin and Elhannan Orren (Tel Aviv, 1982).

Israel State Archives / Central Zionist Archives, *Te'udot Mediniot Ve'diplomatiyot, December 1947–May 1948* (Political and Diplomatic Documents, December 1947–May 1948), ed. Gedalia Yogev (Jerusalem, 1980).

Israel State Archives, *Te'udot Lemediniut Hahutz shel Medinat Yisrael, 14 May–30 September 1948* (Documents on the Foreign Policy of the State of Israel, May–September 1948), vol. i, ed. Yehoshua Freundlich (Jerusalem, 1981).

Israel State Archives, *Te'udot Lemediniut Hahutz shel Medinat Yisrael, October 1948–April 1949* (Documents on the Foreign Policy of the State of Israel, October 1948–April 1949), vol. ii, ed. Yehoshua Freundlich (Jerusalem, 1984).

Israel State Archives, *Te'udot Lemediniut Hahutz shel Medinat Yisrael, Sihot Shvitat–Haneshek im Medinot Arav, December 1948–July 1949* (Documents on the Foreign Policy of the State of Israel, Armistice Negotiations with the Arab States, December 1948–July 1949), vol. iii, ed. Yemima Rosenthal (Jerusalem, 1983).

Israel State Archives, *Te'udot Lemediniut Hahutz shel Medinat Yisrael, May–December 1949* (Documents on the Foreign Policy of Israel, May–December 1949), vol. iv, ed. Yemima Rosenthal (Jerusalem, 1986).

WEITZ, YOSEF, *Yomani Ve'igrotai Labanim* (My Diary and Letters to the Children), vols. iii and iv (Tel Aviv, 1965).

Selected Secondary works

ASSAF, MICHAEL, *Toldot Hit'orerut Ha'aravim Be'eretz Yisrael U'brihatam* (The History of the Awakening of the Arabs in Palestine and Their Flight) (Tel Aviv, 1967).

AVIDAR, YOSEF, 'Tochnit Dalet (Plan D)', in *Sifra Veseifa* no. 2, June 1978.

BANAI (Mazal), Ya'acov, *Hayalim Almonim, Sefer Mivtze'ei Lehi* (Unknown Soldiers, the Book of LHI Operations) (Tel Aviv, 1958).

Be'einei Oyev, Shelosha Pirsumim Arvi'im al Milhemet Hakomemiut (In Enemy Eyes, three Arab publications on the War of Independence), Israel Defence Forces, General Staff/History Branch (Tel Aviv, 1954). The book, translated into Hebrew by Captain S. Sabag, is composed of lengthy excerpts from Mohammed Nimr al Khatib, *Min Athar al Nakba*, Kamal Ismail al Sharif, *Al Ihwan al Muslemin fi Harb Falastin*, and Mohammed Ghussan, *Ma'arak Bab al Wad*.

BEGIN, MENACHEM, *The Revolt* (London, 1964).

CARMEL, MOSHE, *Ma'arachot Tzafon* (Northern Battles) (Tel Aviv, 1949).

Carta's Atlas of Palestine from Zionism to Statehood, ed. Jehuda Wallach (Jerusalem, 1972, 1974).

Carta's Historical Atlas of Israel, The First Years 1948–1961, eds. Jehuda Wallach and Moshe Lissak (Jerusalem, 1978).

COHEN, AHARON, *Yisrael Ve'ha'olam Ha'aravi* (Israel and the Arab World) (Tel Aviv, 1964).

ESHEL, ZADOK, *Hativat Carmeli Bemilhemet Hakomemiut* (The Carmeli Brigade in the War of Independence) (Tel Aviv, 1973).

—— *Ma'arachot Hahaganah Be'Haifa* (The Haganah Battles in Haifa) (Tel Aviv, 1978).

GABBAY, RONY, *A Political Study of the Arab–Jewish Conflict: The Arab Refugee Problem (a Case Study)* (Geneva, Librairie E. Droz, and Paris, Librairie Minard, 1959).

GLUBB, SIR JOHN, *A Soldier with the Arabs* (London, 1957).

Hativat Hanegev Bama'arachah (The Negev Brigade during the War) (Tel Aviv, c.1950).

Hativat Givati Bemilhemet Hakomemiut (The Givati Brigade in the War of Independence), ed. and compiled Major Avraham Eilon (Tel Aviv, 1959).

Hativat Givati mul Hapolesh Hamitzri (The Givati Brigade Opposite the Egyptian Invader), ed. Major Avraham Eilon (Tel Aviv, 1963).

Hativat Alexandroni Bemilhemet Hakomemiut (The Alexandroni Brigade in the War of Independence), Gershon Rivlin and Zvi Sinai (Tel Aviv, 1964).

Ilan Va'shelah, Derekh Hakravot shel Hativat Golani (Tree and Sword, the Route of Battle of the Golani Brigade), ed. and compiled Binyamin Etzioni (Tel Aviv, *c*.1951).

ISRAEL DEFENCE FORCES, GENERAL STAFF HISTORY BRANCH, *Toldot Milhemet Hakomemiut* (History of the War of Independence) (Tel Aviv, 1959).

ITZCHAKI, ARIEH, *Latrun, Hama'aracha al Haderekh Leyerushalayim* (Latrun, the Battle for the Road to Jerusalem), 2 vols. (Jerusalem, 1982).

KAMEN, CHARLES, '*Aharei Ha'ason: Ha'aravim Be'medinat Yisrael, 1948–1950*', ('After the Catastrophe: the Arabs in the State of Israel'), in *Mahbarot Lemehkar U'lebikoret* (Notebooks on Research and Criticism), no. 10, 1985.

KIMCHE, JON and DAVID, *Both Sides of the Hill* (London, 1960).

LAZAR (LITAI), HAIM, *Kibush Yaffo* (The Conquest of Jaffa) (Tel Aviv, 1951).

LORCH, NETANEL, *The Edge of the Sword, Israel's War of Independence, 1947–1949*, revised edn. (Jerusalem, 1968).

Mishmar Ha'emek Ba'ma'aracha (Mishmar Ha'emek in the War) (Tel Aviv, 1950).

MORRIS, BENNY, *The Birth of the Palestinian Refugee Problem, 1947–1949* (Cambridge, 1988).

NAZZAL, NAFEZ, *The Palestinian Exodus from Galilee, 1948* (Beirut, 1978).

NIV, DAVID, *Ma'arachot Ha'irgun Hatz'va'i Ha'leumi* (Battles of the IZL), 6 vols. (Tel Aviv, 1980).

ORREN, ELHANNAN, *Baderekh el Ha'ir, Mivtza Dani* (On the Road to the City, Operation Dani) (Tel Aviv, 1976).

PERETZ, DON, *Israel and the Palestine Arabs* (Washington, DC, 1958).

PORATH, YEHOSHUA, *The Emergence of the Palestinian–Arab National Movement 1918–1929* (London, 1974).

—— *The Palestinian Arab National Movement 1929–1939* (London, 1977).

Sefer Toldot Hahaganah (The History of the Haganah), 3 vols. sub-divided into eight books, eds. Shaul Avigur, Yitzhak Ben-Zvi, Elazar Galili, Yehuda Slutzky, Ben-Zion Dinur, and Gershon Rivlin (Tel Aviv, 1954).

Sefer Hapalmah (The Book of the Palmah), 2 vols., ed. and compiled Zerubavel Gilad together with Mati Megged (Tel Aviv, 1956).

SEGEV, TOM, *1949 Ha'yisraelim Harishonim* (1949, The First Israelis) (Jerusalem, 1984).

SHIMONI, YA'ACOV, *Arviyei Eretz Yisrael* (The Arabs of Palestine) (Tel Aviv, 1947).

SHOUFANI, ELIAS, 'The Fall of a Village', *Journal of Palestine Studies*, 1/4 (1972).

Teveth, Shabtai, *Ben-Gurion and the Palestine Arabs* (Oxford, 1985).

INDEX